Exploring the Dynamic Earth

GIS Investigations for the Earth Sciences

Michelle K. Hall-Wallace

C. Scott Walker

Larry P. Kendall

Christian J. Schaller

Robert F. Butler

The University of Arizona

BROOKS/COLE

THOMSON LEARNING

Australia • Canada • Mexico • Singapore • Spain • United Kingdom • United States

BROOKS/COLE

THOMSON LEARNING

Editor: *Keith Dodson*
Assistant Editor: *Carol Ann Benedict*
Editorial Assistant: *Heidi Blobaum*
Marketing Manager: *Ann Caven*
Advertising Project Manager: *Laura Hubrich*
Project Manager, Editorial Production: *Tom Novack*

Print/Media Buyer: *Kristine Waller*
Permissions Editor: *Sue Ewing*
Cover Designer: *Denise Davidson*
Cover Image: *Digital Vision*
Cover Printer: *Transcontinental*
Printer: *Transcontinental*

For more information about our products, contact us at:
Thomson Learning Academic Resource Center
1-800-423-0563

For permission to use material from this text, contact us by:
Phone: 1-800-730-2214 **Fax:** 1-800-730-2215
Web: http://www.thomsonrights.com

All products used herein are used for identification
purposes only and may be trademarks or registered
trademarks of their respective owners.

Maps and screen shots include data from ESRI Data
and Maps. Data Copyright © ESRI 2002.

ESRI and ArcView are registered trademarks in the United
States and are either trademarks or registered trademarks in
all other countries in which they are used. The ArcView
logo is a trademark of Environmental Systems Research
Institute, Inc.

Earth surface image data used in the Earth globe
illustrations on pages 5 and 104 are © 2001,
TerraMetrics, Inc. Used with permission.

Development of these materials was supported,
in part, by the National Science Foundation under
Grant No. DUE-9555205. Any opinions, findings,
and conclusions or recommendations expressed in
these materials are those of the authors and do not
necessarily reflect the views of the National Science
Foundation.

Library of Congress Control Number: 2002102057

ISBN: 0-534-39138-9

Brooks/Cole—Thomson Learning
511 Forest Lodge Road
Pacific Grove, CA 93950
USA

Asia
Thomson Learning
5 Shenton Way #01-01
UIC Building
Singapore 068808

Australia
Nelson Thomson Learning
102 Dodds Street
South Melbourne, Victoria 3205
Australia

Canada
Nelson Thomson Learning
1120 Birchmount Road
Toronto, Ontario M1K 5G4
Canada

Europe/Middle East/Africa
Thomson Learning
High Holborn House
50/51 Bedford Row
London WC1R 4LR
United Kingdom

Latin America
Thomson Learning
Seneca, 53
Colonia Polanco
11560 Mexico D.F.
Mexico

Spain
Paraninfo Thomson Learning
Calle/Magallanes, 25
28015 Madrid, Spain

Acknowledgments

The authors wish to thank the many students, teachers, and scientists who, through their use of these materials, provided critical reviews and helped us develop insight into how GIS can be most effectively used as a learning and teaching tool.

A significant number of people contributed directly or indirectly to the development of this module, but a few were especially notable. Particular thanks go to Terry Wallace, Joshua Hall, Christine Donovan, Tekla Cook, and Richard Spitzer who tested the investigations multiple times in their classrooms and tirelessly worked with us to improve their content and design. Terry Wallace and Joseph Watkins provided content reviews while Peter Kresan, Carla McAuliffe, and Jo Dodds performed insightful reviews of the pedagogy, design, and content level of the materials. We also appreciate the considerable efforts of our student assistants Marie Renwald, Anne Kramer Huth, Tammy Baldwin, Sara McNamara, Megan Sayles, and Christine Hallman.

We are indebted to the numerous scientists who took the time to learn about our project and share critical research data or expertise that added greatly to the quality of the investigations. Finally, we are grateful to the agencies and individuals that have given us permission to include their outstanding illustrations and photos.

The SAGUARO Project
Michelle K. Hall-Wallace, Director
Department of Geosciences
The University of Arizona
1040 E Fourth Street • Tucson, AZ 85721-0077
saguaro@geo.arizona.edu

Science And GIS Unlocking Analysis & Research Opportunities
http://saguaro.geo.arizona.edu

ESRI Software License Agreement

This is a license agreement and not an agreement for sale. This license agreement (Agreement) is between the end user (Licensee) and Environmental Systems Research Institute, Inc. (ESRI), and gives Licensee certain limited rights to use the proprietary ESRI® desktop software and software updates, sample data, online and/or hard-copy documentation and user guides, including updates thereto, and software keycode or hardware key, as applicable (hereinafter referred to as "Software, Data, and Related Materials"). All rights not specifically granted in this Agreement are reserved to ESRI.

Reservation of Ownership and Grant of License: ESRI and its third party licensor(s) retain exclusive rights, title, and ownership of the copy of the Software, Data, and Related Materials licensed under this Agreement and, hereby, grants to Licensee a personal, nonexclusive, nontransferable license to use the Software, Data, and Related Materials based on the terms and conditions of this Agreement. From the date of receipt, Licensee agrees to use reasonable effort to protect the Software, Data, and Related Materials from unauthorized use, reproduction, distribution, or publication.

Copyright: The Software, Data, and Related Materials are owned by ESRI and its third party licensor(s) and are protected by United States copyright laws and applicable international laws, treaties, and/or conventions. Licensee agrees not to export the Software, Data, and Related Materials into a country that does not have copyright laws that will protect ESRI's proprietary rights. Licensee may claim copyright ownership in the Simple Macro Language (SML™) macros, AutoLISP® scripts, AtlasWare™ scripts, and/or Avenue™ scripts developed by Licensee using the respective macro and/or scripting language.

Permitted Uses:
- Licensee may use the number of copies of the Software, Data, and Related Materials for which license fees have been paid on the computer system(s) and/or specific computer network(s) for Licensee's own internal use. Licensee may use the Software, Data, and Related Materials as a map/data server engine in an Internet and/or Intranet distributed computing network or environment provided the appropriate, additional license fees are paid. If the Software, Data, and Related Materials contain dual media (i.e., both 3.5-inch diskettes and CD–ROM), then Licensee may only use one (1) set of the dual media provided. Licensee may not use the other media on another computer system(s) and/or specific computer network(s), or loan, rent, lease, or transfer the other media to another user.
- Licensee may install the number of copies of the Software, Data, and Related Materials for which license or update fees have been paid onto the permanent storage device(s) on the computer system(s) and/or specific computer network(s).
- Licensee may make routine computer backups but only one (1) copy of the Software, Data, and Related Materials for archival purposes during the term of this Agreement unless the right to make additional copies is granted to Licensee in writing by ESRI.
- Licensee may use, copy, alter, modify, merge, reproduce, and/or create derivative works of the online documentation for Licensee's own internal use. The portions of the online documentation merged with other software, hard copy, and/or digital materials shall continue to be subject to the terms and conditions of this Agreement and shall provide the following copyright attribution notice acknowledging ESRI's proprietary rights in the online documentation: "Portions of this document include intellectual property of ESRI and are used herein by permission. Copyright © 200_ Environmental Systems Research Institute, Inc. All Rights Reserved."
- Licensee may use the Data that are provided under license from ESRI and its third party licensor(s) as described in the Distribution Rights section of the online Data Help files.

Uses Not Permitted:
- Licensee shall not sell, rent, lease, sublicense, lend, assign, time-share, or transfer, in whole or in part, or provide unlicensed third parties access to prior or present versions of the Software, Data, and Related Materials, any updates, or Licensee's rights under this Agreement.
- Licensee shall not reverse engineer, decompile, or disassemble the Software, or make any attempt to unlock or bypass the software keycode and/or hardware key used, as applicable, subject to local law.
- Licensee shall not make additional copies of the Software, Data, and/or Related Materials beyond that described in the Permitted Uses section above.
- Licensee shall not remove or obscure any ESRI copyright or trademark notices.
- Licensee shall not use this software for more than one hundred twenty (120) days from the date that the software was installed. At the end of this period, users must remove the time limited software from their computers or purchase fully licensed software. Students and instructors in the United States may purchase fully licensed individual copies of the software from ESRI telesales at 1-800-GIS-XPRT.

Term: The license granted by this Agreement shall commence upon Licensee's receipt of the Software, Data, and Related Materials and shall continue until such time that (1) Licensee elects to discontinue use of the Software, Data, and Related Materials and terminates this Agreement or (2) ESRI terminates for Licensee's material breach of this Agreement. Upon termination of this Agreement in either instance, Licensee shall return to ESRI the Software, Data, Related Materials, and any whole or partial copies, codes, modifications, and merged portions in any form. The parties hereby agree that all provisions, which operate to protect the rights of ESRI, shall remain in force should breach occur.

Limited Warranty: ESRI warrants that the media upon which the Software, Data, and Related Materials are provided will be free from defects in materials and workmanship under normal use and service for a period of sixty (60) days from the date of receipt. The Data herein have been obtained from sources believed to be reliable, but its accuracy and completeness, and the opinions based thereon, are not guaranteed. Every effort has been made to provide accurate Data in this package. The Licensee acknowledges that the Data may contain some nonconformities, defects, errors, and/or omissions. ESRI and third party licensor(s) do not warrant that the Data will meet Licensee's needs or expectations, that the use of the Data will be uninterrupted, or that all nonconformities can or will be corrected. ESRI and the respective third party licensor(s) are not inviting reliance on these Data, and Licensee should always verify actual map data and information. The Data contained in this package are subject to change without notice.

EXCEPT FOR THE ABOVE EXPRESS LIMITED WARRANTIES, THE SOFTWARE, DATA, AND RELATED MATERIALS CONTAINED THEREIN ARE PROVIDED "AS IS," WITHOUT WARRANTY OF ANY KIND, EITHER EXPRESS OR IMPLIED, INCLUDING, BUT NOT LIMITED TO, THE IMPLIED WARRANTIES OF MERCHANTABILITY AND FITNESS FOR A PARTICULAR PURPOSE.

Exclusive Remedy and Limitation of Liability: During the warranty period, ESRI's entire liability and Licensee's exclusive remedy shall be the return of the license fee paid for the Software, Data, and Related Materials in accordance with the ESRI Customer Assurance Program for the Software, Data, and Related Materials that do not meet ESRI's Limited Warranty and that are returned to ESRI or its dealers with a copy of Licensee's proof of payment.

ESRI shall not be liable for indirect, special, incidental, or consequential damages related to Licensee's use of the Software, Data, and Related Materials, even if ESRI is advised of the possibility of such damage.

Waivers: No failure or delay by ESRI in enforcing any right or remedy under this Agreement shall be construed as a waiver of any future or other exercise of such right or remedy by ESRI.

Order of Precedence: Any conflict and/or inconsistency between the terms of this Agreement and any FAR, DFAR, purchase order, or other terms shall be resolved in favor of the terms expressed in this Agreement, subject to the U.S. Government's minimum rights unless agreed otherwise.

Export Regulations: Licensee acknowledges that this Agreement and the performance thereof are subject to compliance with any and all applicable United States laws, regulations, or orders relating to the export of computer software or know-how relating thereto. ESRI Software, Data, and Related Materials have been determined to be Technical Data under United States export laws. Licensee agrees to comply with all laws, regulations, and orders of the United States in regard to any export of such Technical Data. Licensee agrees not to disclose or reexport any Technical Data received under this Agreement in or to any countries for which the United States Government requires an export license or other supporting documentation at the time of export or transfer, unless Licensee has obtained prior written authorization from ESRI and the U.S. Office of Export Control. The countries restricted at the time of this Agreement are Cuba, Iran, Iraq, Libya, North Korea, Serbia, and Sudan.

U.S. Government Restricted/Limited Rights: Any software, documentation, and/or data delivered hereunder is subject to the terms of the License Agreement. In no event shall the Government acquire greater than RESTRICTED/LIMITED RIGHTS. At a minimum, use, duplication, or disclosure by the Government is subject to restrictions as set forth in FAR §52.227-14 Alternates I, II, and III (JUN 1987); FAR §52.227-19 (JUN 1987) and/or FAR §12.211/12.212 (Commercial Technical Data/Computer Software); and DFARS §252.227-7015 (NOV 1995) (Technical Data) and/or DFARS §227.7202 (Computer Software), as applicable. Contractor/Manufacturer is Environmental Systems Research Institute, Inc., 380 New York Street, Redlands, CA 92373-8100, USA.

Governing Law: This Agreement is governed by the laws of the United States of America and the State of California without reference to conflict of laws principles.

Entire Agreement: The parties agree that this constitutes the sole and entire agreement of the parties as to the matter set forth herein and supersedes any previous agreements, understandings, and arrangements between the parties relating hereto and is effective, valid, and binding upon the parties.

ESRI is a trademark of Environmental Systems Research Institute, Inc., registered in the United States and certain other countries; registration is pending in the European Community. SML, AtlasWare, and Avenue are trademarks of Environmental Systems Research Institute, Inc.

Introduction

Getting started

Unit 1 – Searching for Evidence

Unit 2 – Exploring Plate Tectonics

Unit 3 – Earthquake Hazards

Unit 4 – Volcano Hazards

Unit 5 – Tsunami Hazards

Introduction

Thinking scientifically

A geoscientist makes a living by observing and measuring nature. Whether recording a violent volcanic eruption or mapping the trace of a fault, a successful Earth scientist relies heavily on his or her ability to recognize patterns. Patterns in space and time are the keys to many of the great discoveries about how the Earth works. The activities in this module will help you develop your ability to recognize and interpret nature's fundamental patterns by exploring modern scientific data using a geographic information system (GIS). A GIS is a tool for organizing, manipulating, analyzing, and visualizing information about the world using a computer.

Through these exercises you will examine the global patterns of earthquakes, volcanoes and topography found around the world, retracing some of the same analytical steps followed by scientists as they gradually uncovered the process of plate tectonics. Armed with seafloor age data and your knowledge of plate tectonics, you will peer into Earth's past to learn about the ages of the Atlantic and Pacific Oceans, and gaze into a future in which Los Angeles and San Francisco are no longer rivals, but suburbs of one another! Moving from geologic to human time frames, you will examine some recent geologic events, including earthquakes, volcanic eruptions and tsunamis, to evaluate the risk to societies worldwide from future occurrences of these geologic hazards.

Most of these patterns are presented in maps, which are one of the geologist's most important tools. The maps will allow you to explore the relationships between natural features and phenomena such as the location of rivers, mountains, earthquakes and volcanic eruptions and human features such as roads, cities, and population density. By investigating a map of population density and locations of deadly earthquakes, you can use the patterns in the data to deduce the nature of the relationship between these features. Maps also provide a convenient way to present statistical relationships, and maps of different time intervals allow you to look at changes in features over time. You will even be able to create your own maps and explore!

A GIS Map

This map shows earthquakes that have occurred in parts of North America and Central America as well as the northern tip of South America.

Tabular Data

A tabular form of the same earthquake data. The yellow row shows a selected earthquake.

Planning to learn

Each unit will take you through a well-tested learning process that helps you examine your existing knowledge and build upon it. The first activity will get you thinking about your present knowledge of the major concepts in the unit. It may stimulate questions that you have about the topic. Write these down and, as you learn more, see if you can answer them for yourself. In the second activity you will explore maps and data looking for patterns.

When exploring a particular feature, ask questions such as the following:

- Where do they occur?
- Where don't they occur?
- Why is it there and not elsewhere?
- Is it distributed regularly or irregularly?
- What might cause this pattern?

The third activity provides readings with key information about the major concepts and should help you begin to answer the questions raised above. The readings have been kept brief and provide only the important points you need to continue on or check your prior answers. In the fourth activity, you will apply your new knowledge to solve a particular problem. This will help you measure your understanding of the material and prepare you for any assessment later on.

GIS made easier

The purpose of these activities is not to learn how to use a GIS, but to use one as a tool to explore and learn about natural processes and features and how they relate to humans and human activities. For this reason, all of the data have been assembled into ready-to-use projects and complex operations have been eliminated or simplified. While it's good to know basic computer skills, you don't need experience with ArcView GIS software to do these activities. Directions for each new task are provided in the text, so you will learn to use the tool as you explore with it. It is especially important to pay attention to the tips provided in the margin, such as the *Want to know more?* item at the left. The activities barely scratch the surface of the data that have been provided, and we encourage you to explore them on your own and make your own discoveries.

Using these materials

Visual cues are used to make the activity directions easier to follow.

- A line preceded by the ▶ symbol is an instruction—something to do on the computer.
- When referring to a tool or button, the name of the tool or button is capitalized and is followed by a picture of that item as it appears on screen—e.g. …the Identify tool ⊙.
- The ▶ symbol between two boldface words in text indicates a menu choice. Thus, **Theme ▶ Properties** means "pull down the Theme menu and choose Properties."

The most common mistake made when using ArcView GIS software is not activating the correct theme (map layer) before performing an operation. When things don't seem to be going as they should, this is the first thing to check.

The second most common problem is not looking closely enough at the maps to see what's going on. ArcView has several tools and buttons for zooming in and out of the map view that work just like tools you've used in other applications—use them!

Want to know more?

If you would like more information on how to use ArcView GIS, refer to the *Guide to ArcView GIS*, found in the **Docs** folder on the CD.

Theme ▶ Properties *means...*

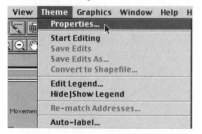

To activate a theme, click on its name in the Table of Contents. Active themes are indicated by a raised border.

To zoom in on an area, click and drag with the Zoom In tool ⊕ to outline the area. **To zoom out**, click anywhere on the map window with the Zoom Out tool ⊖.

Getting started

Additional resources

Visit The SAGUARO Project website for updates, references, and links to related websites:

http://saguaro.geo.arizona.edu

The Exploring the Dynamic Earth CD contains all the software and data you need to complete the module activities on your own Macintosh® or Windows® computer. If you are using this manual as part of a laboratory course, a computer lab with the necessary software and data files may have already been prepared for you in advance.

What you need to know

The authors of this book assume that you know how to use a computer with either the Macintosh or Windows operating system installed. We will make no attempt to teach these basic skills:

Installation courtesy

Install the Exploring the Dynamic Earth applications and data only on your own personal computer. If someone else owns the machine, you should seek permission before installing anything.

- turning on the computer and, if necessary, logging in as a user;
- navigating the computer's file system to find folders, applications, and files;
- launching applications and opening files from within those applications; and
- using basic interface elements—opening, closing, moving, and resizing windows, using tools, menus, and dialog boxes, etc.

The data files for this module may be accessed directly from the CD or copied to a user-specified drive and directory during the installation process. Thus, in our modules, you will be instructed to launch the ArcView application, then locate and open a specific file. In a lab setting, your instructor will tell you where to find this file.

Minimum system requirements

Macintosh

Mac OS X compatibility

ArcView GIS 3.0a for Macintosh runs in Classic mode under Mac OS X. For best results, restart your computer in OS 9 and install ArcView GIS.

- ArcView® GIS 3.0a for Macintosh®
- QuickTime™ and Acrobat® Reader (installers included on CD)
- 133 MHz or faster PowerPC® CPU running Mac OS 8.0 or newer
- 64 MB total RAM (32 MB of available application RAM)
- CD-ROM drive
- 40 MB available hard disk space (+ 100 MB for data, if installed on hard disk)

Windows

- ArcView® GIS 3.0a or higher for Windows®
- QuickTime™ and Acrobat® Reader (installers included on CD)
- 133 MHz or faster Pentium™-class CPU running Windows 95 or newer
- 64 MB total RAM (32 MB of available application RAM)
- CD-ROM drive
- 40 MB available hard disk space (+ 100 MB for data, if installed on hard disk)

Macintosh software installation

The instructions below are for installing software on your personal Macintosh computer. If the software has already been installed, such as in a lab setting, skip ahead to the **Using ArcView GIS for Macintosh** section on page vii.

Before you install software

- Insert the **Exploring the Dynamic Earth** CD into your CD-ROM drive and read the **ReadMe** file on the CD for installation updates.
- Disable virus protection software (if installed) and quit any open applications.
- You must have at least 40 MB of free space on your hard drive to install the application. Installing the module data on your hard drive is optional and requires an additional 100 MB of drive space.

Installing ArcView

- Open the **ArcView** folder, double-click the **ArcView Installer** icon, and follow the on-screen instructions.
- When installation has finished, restart your computer and hold down the Command (⌘) and Option keys until the message "Are you sure you want to rebuild the desktop file on the disk..." message appears. Click **OK**. Your computer will rebuild its desktop file and complete the startup process.

Setting ArcView's memory allocation

The default memory allocation for ArcView must be increased to assure trouble-free operation! Follow these steps to increase the application's memory allocation.

- Navigate to where you installed the ArcView GIS application. Look for a folder named **ESRI**—open it and open the folder inside it named **AV_GIS30a**.
- Single-click the **ArcView** application icon 🔲 to select it.
- Choose **File ▶ Get Info**, choose **Memory** from the **Show** pop-up menu.
- Enter 32000 for the **Preferred Size** and at least 32000 for the **Minimum Size. (Note—don't include commas!)**

Macintosh CD-ROM Contents

The **Exploring the Dynamic Earth CD** contains the following folders and files.

Readme.txt
License.txt
Dynamic_Earth
  dynamic.apr
  hazards.apr
  sagemdia.txt
  Data (many files)
  Media (many files)
ArcView
  ArcView Installer
Reader
  Acrobat Reader Installer
QuickTime
  QuickTime Installer
Docs
  Guide to ArcView GIS.pdf
  Data Dictionary.pdf

"Can I allocate more memory to ArcView?"

If your computer has more memory available, you can allocate some of the additional memory to ArcView, keeping in mind your other system requirements.

```
Memory Requirements

Suggested Size:   12004   K
Minimum Size:    [32000]  K
Preferred Size:  [48000]  K
```

Where is ArcView GIS?

On a Macintosh, ArcView GIS is installed in a folder named **ESRI** on the drive you specified during installation. You will find the ArcView application in the **AV_GIS30a** folder.

Make an alias!

For convenience, you may wish to make an alias of the ArcView application on your desktop. See your documentation for instructions on how to create an alias.

Monitor resolution

The Exploring the Dynamic Earth activities were designed to be used on a computer with a monitor resolution of at least 800 by 600 pixels and 256 colors. Be sure to set your monitor accordingly, if necessary.

Changing monitor resolution on a Macintosh computer

Select Monitors from the Control Panels under the Apple menu. (Your particular computer may say "Monitors and Sound.") Select the Monitor button and choose the appropriate screen resolution.

Registering ArcView GIS

To register the ArcView GIS application (Macintosh only):

- Double-click the **ArcView** icon 🔍 to launch the application.
- When prompted, enter your name and company (you may leave this blank), then enter the registration number: **708301184404**.
- Quit ArcView.

Installing QuickTime

The investigations in this module use QuickTime for displaying movies and animations. Most Macintosh computers already have QuickTime installed. If yours does not, you should install it.

- Turn on your computer and insert the **Exploring the Dynamic Earth** CD into the CD-ROM drive.
- Open the **QuickTime** folder on the CD and double-click the **Quicktime Installer** icon. Follow the on-screen instructions to complete the installation.
- Restart your computer.

Installing Acrobat Reader

Acrobat Reader is used to view and print the files in the Docs folder. If Acrobat Reader is already installed on your computer or you do not wish to view or print these documents, you may skip this installation.

- Turn on your computer and insert the **Exploring the Dynamic Earth** CD into the CD-ROM drive.
- Open the **Reader** folder on the CD and double-click the **Acrobat Reader Installer** icon. Follow the on-screen instructions to complete the installation.
- Restart your computer.

Using ArcView GIS for Macintosh

Launching ArcView and opening project files

- Double-click the **ArcView** application icon 🔍 to launch ArcView.
- Choose **File ▶ Open**.
- Navigate to the **Dynamic_Earth** folder (on your **Exploring the Dynamic Earth** CD or installed on your hard drive), select the ArcView project file **dynamic.apr** or **hazards.apr**, and click **Open**. ArcView project files end in ".apr."

Closing project files

When you have completed an activity or must stop for some reason,

- Choose **File ▶ Quit**.
- When asked if you want to save your changes, click **No**. (Don't worry if you click **Yes**. The files have been locked to prevent accidentally modifying or erasing them.)

Windows software installation

The instructions below are for installing software on your personal Windows-based computer. If the software has already been installed, such as in a lab setting, skip ahead to the **Using ArcView GIS for Windows** section on page ix.

Before you install software

- Disable virus protection software (if installed) and quit any open applications.
- Make sure you have sufficient hard disk space for the installation—40 MB for the application. Installing the module data on your hard disk is optional and requires approximately 100 MB of additional hard disk space.
- You may install this version of ArcView alongside older or newer versions by installing it in a different location on your hard disk.

Installing ArcView GIS

- NOTE—This is a 120-day license. Do not install until you need to use it, or your license may expire before the semester is over!
- Insert the **Exploring the Dynamic Earth** CD into your CD-ROM drive.
- Open the **ArcView** folder, double-click the **AV32AVC.EXE** icon, and follow the on-screen instructions to complete the installation. *(Note - newer versions of Windows may be configured to hide the three character file extension, so this file would appear as simply AV32AVC.)*
- Restart your computer.

Installing QuickTime for Windows

This module uses QuickTime for displaying animations and movies. To install the QuickTime Player application:

- Turn on your computer and insert the **Exploring the Dynamic Earth** CD into the CD-ROM drive.
- Open the **QuickTime** folder on the CD and double-click the **QuickTimeInstaller.exe** icon. Follow the on-screen instructions to complete the installation.
- Restart your computer.

Installing Acrobat Reader for Windows

Acrobat Reader is used to view and print the files in the Docs folder. If Acrobat Reader is already installed on your computer or you do not wish to view or print these documents, you may skip this installation.

- Turn on your computer and insert the **Exploring the Dynamic Earth** CD into the CD-ROM drive.
- Open the **Reader** folder on the CD and double-click the **RP505ENU.EXE** icon. Follow the on-screen instructions to complete the installation.
- Restart your computer.

Windows CD-ROM Contents

The Exploring the Dynamic Earth CD-ROM contains the following folders and files.

Readme.txt
License.txt
Dynamic_Earth
  dynamic.apr
  hazards.apr
  sagemdia.txt
  Data (many files)
  Media (many files)
ArcView
  AV32AVC.EXE
Reader
  RP505ENU.EXE
QuickTime
  QuickTimeInstaller.exe
Docs
  Guide to ArcView GIS.pdf
  Data Dictionary.pdf

Monitor resolution

The Exploring the Dynamic Earth module was designed to be used with a monitor resolution of at least 800 by 600 pixels and 256 colors.

Changing monitor resolution on a Windows computer

Right-click on the desktop, choose Settings from the popup menu, and click the Settings tab. Set the color palette and desktop area to appropriate values.

For more help, consult your computer's printed or online documentation.

Using ArcView GIS for Windows

Launching ArcView GIS for Windows

- On the Windows desktop, click the **Start** button.
- On the **Start** menu, choose **Programs ▶ ESRI ▶ ArcView GIS Virtual Campus Edition ▶ ArcView GIS Virtual Campus Edition**.
- On the next screen, click the **Try** button. This will begin your 120-day license period. Each time you launch the ArcView GIS application, this screen tells you the number of days remaining in the license period.

Create a shortcut

You may want to create a shortcut to ArcView on your desktop, to make it easier to access the program. To find out how to create a shortcut, refer to your Windows documentation.

Time remaining on 120-day license

Click here to start using ArcView

Opening a project file

- Launch ArcView GIS for Windows (see above).
- In the **Welcome to ArcView GIS** dialog box, click the **Open an existing project** button, then click **OK**.
- Use the **Open** dialog box to locate the file you want to open:
  1) Select the drive where you installed the module files.
  2) Select the directory (folder) containing the project files.
  3) Select the file. ArcView project files end in ".apr."
  4) Click **OK**.

Where is ArcView?

The installer places ArcView in the **Program Files** folder of the drive you specified, in a folder named **VCampus**.

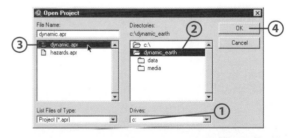

- When the project files have opened, you will see the message shown at left. Answer either "yes" or "no"—it makes no difference.

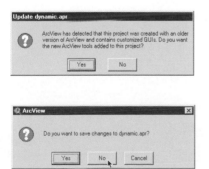

Closing a project file

When you have completed an activity or must stop for some reason,

- Choose **File ▶ Quit**.
- When asked if you want to save your changes, click **No**. (Don't worry if you click **Yes**. The files have been locked to prevent accidentally modifying or erasing them.)

Accessing the module data

If ArcView and QuickTime software are already installed on your computer, you are ready to use the module. You may access the module data directly from the CD-ROM or copy it to your computer's hard drive for increased convenience and performance.

Reading the data files directly from the CD

- Insert the **Exploring the Dynamic Earth** CD in your CD-ROM drive.
- Launch ArcView, choose File ▶ Open, navigate to the **Dynamic_Earth** folder on the CD, and open the desired ArcView project file.

Copying the data files to your computer

- Insert the **Exploring the Dynamic Earth** CD in your CD-ROM drive.
- Copy the *entire* **Dynamic_Earth** folder from the CD to your hard disk.
- **Important** - Do not change the name of the **Dynamic_Earth** folder or any of its contents at any time.
- **Important** - If you are copying the **Dynamic_Earth** folder inside another folder, give the folder a short name (8 characters or fewer) and do not nest the folder too many levels deep. This will help ensure that ArcView can locate the files correctly.
- **Critical issue for Windows installations only** - There must be *no spaces* in the names of the drive or folders along the path to the **Dynamic_Earth** folder. If necessary, change spaces in drive and folder names to underscore characters—thus, a **Class Data** folder should be renamed **Class_Data**.
- If the data files get renamed or damaged, delete the entire **Dynamic_Earth** folder and copy a clean version of the folder from the CD to your hard disk.

Q&A about the 120-day ArcView license on the CD

"My university has an ArcView site license—do I need to use the 120-day version of ArcView on the Exploring the Dynamic Earth CD?"

Only if you want to use your own computer to complete the GIS activities—the 120-day version of ArcView is primarily intended for use on students' personal computers.

"When should I install the 120-day version of ArcView?"

Do not install your ArcView GIS software until you need to use it. The ArcView license allows you to install the ArcView application on one computer for 120 days. At the end of that time, the software will no longer function, and it cannot be reinstalled on that computer.

"Can I reinstall the application if it gets damaged or someone accidentally uninstalls it?"

Yes, but it will only function for 120 days from the original installation date.

"What happens at the end of the 120-day license period?"

Your copy of ArcView GIS will stop working. According to the license agreement, you must uninstall the ArcView GIS application, since it will no longer function and will not work if reinstalled. For pricing and information about student and instructor licenses for ArcView, call ESRI telesales at (800) 447-9778.

"What happens if I retake this course or take a different course using another module from the Exploring series?"

You must install your "new" 120-day licensed version of ArcView on a different computer than your original installation.

ArcView troubleshooting

To activate a theme, click on its name in the Table of Contents. Active themes are indicated by a raised border.

"When I _____ in ArcView, nothing (or the wrong thing) happens."

Most ArcView errors are caused by not having the correct theme activated when performing an operation. By activating a theme, you are telling ArcView which data to operate on. If the wrong data are identified, the operation will most likely fail or produce unexpected results.

"ArcView crashes while launching." (Macintosh and Windows)

- ArcView may not have enough memory to work properly (Macintosh only)—see page vi for directions on **Setting ArcView's memory allocation**.
- ArcView may be damaged. You may uninstall and reinstall ArcView at any time within the 120-day license period using the **Exploring the Dynamic Earth** CD.

"When I open a project file, ArcView keeps telling me it can't find files that I know are installed." (Macintosh and Windows)

The project file contains a path to the location of each data file in the project. If you move files or rename any of the files or folders that were installed, ArcView cannot find the files. If you cannot restore the correct file and folder names and locations, delete the **Dynamic_Earth** folder and re-install it on the hard disk.

"When I open a project file from within ArcView, it tells me it can't find the project file." (Windows only)

There is probably a space in the name of the drive or folder into which you copied the **Dynamic_Earth** folder. ArcView for Windows cannot locate a file correctly if it encounters a space in the file's pathname.

"ArcView launches and opens the project file, but it crashes later, after several minutes of use." (Mac)

The memory allocation for ArcView is probably too low. See page vi for directions on **Setting ArcView's memory allocation**.

"When I use ArcView's Media Viewer to view movies or animations, I get a message inviting me to upgrade to QuickTime Pro. When I click the "later" button, nothing happens." (Windows only)

This usually occurs only in a lab or other shared environment, when the QuickTime preferences file is not "writable" to all users. The preferences file is called **QuickTime.qtp**; its exact location depends on the version of Windows you are using. Consult your system administrator about making the necessary changes.

Updates and resources

For corrections, updates, and learning tips for this module, visit the SAGUARO Project website at:

http://saguaro.geo.arizona.edu

Online help

Two Help menus on a Mac?

On the Macintosh, ArcView adds a second Help menu of its own to the left of the standard Macintosh Help menu. Use this menu to access ArcView's built-in help system.

Window	Help	Help
	Help Topics...	
	How to Get Help...	
	About ArcView...	

This module provides all of the directions you need to complete the activities using ArcView GIS. If you wish to explore ArcView on your own, or learn more about ArcView's capabilities, you may wish to consult ArcView's **Help** menu. Choose **Help ▶ How to Get Help...** to learn more about using ArcView's built-in help system.

Printing the Educator's Guide to ArcView GIS

The **Docs** folder contains a handy quick reference manual of the most common tools and techniques used in ArcView, called the Educator's Guide to ArcView GIS. You may view this document on screen or print all or part of it for later reference using the included Acrobat Reader software.

- Turn on your computer and insert the **Exploring the Dynamic Earth** CD into the CD-ROM drive.
- Open the **Docs** folder on the CD and double-click the **Guide to ArcView GIS.pdf** icon.
- Use the on-screen controls to scroll through the document, and choose **File ▶ Print** and either print all pages or select a range of pages to print.

The Data Dictionary file

The **Data Dictionary.pdf** file found in the **Docs** folder provides information about the data used in this module, including a description of where the data came from and how they were processed. You do not need this information to complete the activities in the module, but you may find it useful if you are using the data for independent research or just to satisfy your curiosity. Many of the data providers listed are good sources for additional data and information.

Unit 1
Searching for Evidence

In this unit, you will...

- *investigate patterns in seafloor topography, earthquakes, and volcanoes,*

- *examine plate motion data, and*

- *use these patterns and data to identify and classify plate boundaries.*

This view of global topography provides important evidence for the processes shaping Earth's crust.

Activity 1.1

Investigating a changing Earth

As a class, or in small groups as assigned by your instructor, discuss these questions.

1. The map on the first page of this unit shows that Earth is not flat or smooth. What do you think causes variations in Earth's surface?

2. Is the Earth changing? What is the evidence for or against this change on the Earth? List any *observable* evidence you can think of that shows that the surface of our planet is or is not changing. The evidence can include both physical features and processes.

3. How long does it take to observe change or lack of change on Earth?

4. Give examples of features that we can observe changing or remaining the same, and the time over which the process occurs.

In later activities, you will explore data to look for patterns in topography (the shape of the land), volcanoes, earthquakes, and the ages of seafloor rocks. These will provide important clues about how Earth's surface formed.

Activity 1.2

Investigating Earth's clues

Things to know...

The ▶ symbol between two boldface words in text indicates a menu choice. Thus, **File ▶ Open...** means "pull down the File menu and choose Open... from the menu."

How to zoom in and out. Use the Zoom In 🔍 and Zoom Out 🔍 tools. To see the entire map again, choose View > Full Extent.

To turn a theme on or off, click its checkbox in the Table of Contents.

To activate a theme, click on its name in the Table of Contents. Active themes are indicated by a raised border.

In this activity, you will explore patterns of features and events on Earth's surface. You will look for clues to help you understand how these patterns are interrelated and what they tell about the history of our planet. Your main tool in these investigations is a computer application called ArcView GIS, a Geographic Information System. To get started, you only need to know how to do a few basic things with ArcView (see note at left). You will learn many new skills along the way.

▶ Launch the ArcView GIS application, then locate and open the **dynamic.apr** project file.

▶ From the list of views, open the **Clues** view. It should look like this:

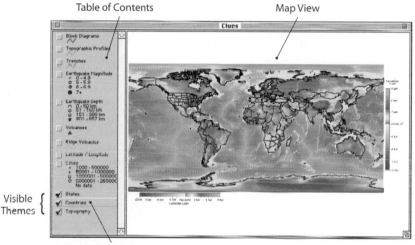

Topography tales

Inside the Earth

The Earth is composed of distinct layers with the densest material in the core and the least dense material in the thin crust.

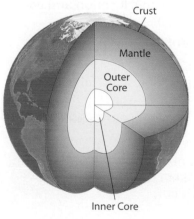

Earth's story is told through its surface features, or **topography**. If Earth were made of only one kind of material, there would be no dry land. Earth's surface would be smooth and covered by a single global ocean. Fortunately, our planet is made of many types of rock, each with a different density. As gravity pulls the denser rock toward Earth's center, the less dense rock is pushed toward the surface. The result is a broken, bumpy surface with magnificent mountains of low density rocks and deep ocean basins of higher density rocks. You will begin this investigation by looking at Earth's topography in detail.

This map is a color shaded relief map. The colors represent elevations, *not* the surface material or the ground cover.

1. Using the elevation scale t the right or below the map, what color is each of the following?

    a. The lowest topography –

    b. Sea level –

    c. The highest topography –

The highest highs

The obvious place to look for evidence of change is where the topography is most extreme—its highest and lowest places. You will start your search by looking for Earth's highest places.

▶ Turn on and activate the **Topographic Profiles** theme.

Still can't picture it?

To better understand what topographic profiles represent, click the Media Viewer button 🖾 and choose **Topographic Profiles** from the media list.

The lines in this theme mark the locations of *topographic profiles*, or cross-sectional views of the land surface. Imagine slicing through the Earth, pulling the halves apart, and viewing one half from the side. The result is a topographic profile. For example, the topographic profile below shows a slice through Mexico's Yucatán Peninsula. To make the land features easier to see, topographic profiles are usually "stretched" vertically—a process called *vertical exaggeration*.

Profile location map

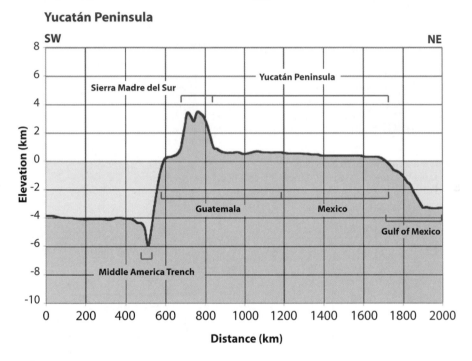

Yucatán Peninsula

▶ Examine the **Clues** view. Using the map elevation scale at the bottom or to the right of the map, find a place where a profile line crosses an area 3 kilometers or higher in elevation.

Next, you will sketch the topographic profile for this location and use it to estimate the highest elevation along the profile.

▶ Using the Hot Link tool ⚡, click the profile line you chose.

2. Sketch and label the topographic profile in the first grid provided on the next page. Include all geographic labels and label the distance on the horizontal (X) axis.

3. Estimate the highest elevation along the profile line and write it in the space provided above the grid.

Hot tips for Hot Links

When you use the Hot Link tool ⚡,

• The theme containing the hot links must be active (the tool on the tool bar will be dark ⚡, not grayed out ⚡).

• Click when the *tip* of the lightning bolt cursor is on the feature.

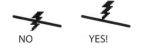

NO   YES!

▶ Close the profile plot window when finished. Locate three more profiles that cross areas higher than 3 km and repeat the above process in the other three grids provided.

Location _____

Maximum elevation = _____ km

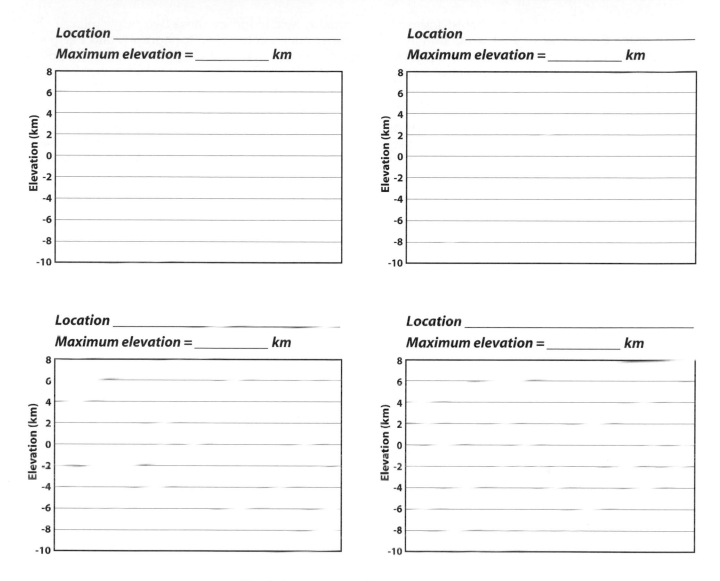

Location _____

Maximum elevation = _____ km

Location _____

Maximum elevation = _____ km

Location _____

Maximum elevation = _____ km

Block diagrams

Think of block diagrams as pieces cut from the middle of a large layer cake and viewed at an angle. Notice that the upper edge of each face of the block is a topographic profile. In addition to surface features, block diagrams usually show layers of rock and structures below the surface.

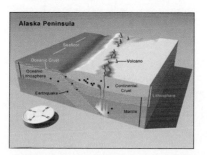

Like icing on a cake

Greenland and Antarctica are two of Earth's highest regions. Much of their elevation, however, is not rock but another material. (Remember, the brown color in the shaded relief image represents high elevation, *not* the surface material!) To compare Greenland and Antarctica with other high regions and find out what this "other material" is, you will examine block diagrams of the areas.

▶ Turn on and activate the **Block Diagrams** theme.

▶ Using the Hot Link tool ⚡, click the profile lines crossing Greenland, Antarctica, and any other high regions you identified above, to display a block diagram of each location. Close each block diagram window when you are finished viewing it.

4. What is the material that adds as much as 1,500 meters (5,000 feet) to the elevation of Greenland and Antarctica?

Not-so-green Greenland & not-so-icy Iceland

Place names can be deceiving. While there's some green land along the coasts of Greenland, 85% of the land is covered by a continental ice sheet up to 3 km thick. This represents 10% of the world's total fresh water reserves.

In contrast, only 11% of neighboring Iceland is covered with glaciers and ice fields, and these are much thinner than Greenland's ice sheet.

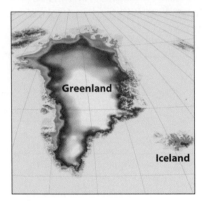

For this reason, it's probably best to ignore these two "high" places when thinking about the connection between topography and Earth's changing surface. Still, even these places fit the pattern of less dense material forming higher topography.

The ocean's unexpected highs

Before maps of seafloor topography became available, most scientists believed the seafloor was relatively flat. Personal experience with lakes and rivers often suggests that the deepest part of the ocean should be near the middle.

5. Examine the ocean basins. Are they smooth bottomed? Does their depth increase continuously toward the middle? If not, describe the features you do find there. (Are these features like plains, like valleys, or like mountains?)

▶ Use the Zoom In tool 🔍 to take a closer look at the seafloor between Florida and the coast of Africa.

▶ Activate the **Topographic Profiles** theme.

▶ Using the Hot Link tool ⚡, click on the profile line that crosses that region.

6. Sketch the topography along the profile line going from Florida to Africa.

▶ Choose **View ▶ Full Extent** to view the entire map again.

Spreading ridge

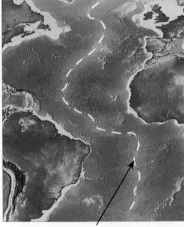

shallow linear ridge in mid-Atlantic Ocean

Features like the one running down the middle of the Atlantic Ocean are called *spreading ridges*. To the eye they look like long, jagged scars and in a sense they are, as you'll see later. To better understand the topography of the Mid-Atlantic Ridge, look at the three topographic profiles that cross it.

Reading coordinates: The cursor's longitude and latitude are displayed on the right end of the tool bar.

113.83 → — longitude
-20.73 ↑ — latitude

▶ Using the Hot Link tool ⚡, click on each of the profile lines that cross the ridge and examine their profiles. These are located at about 23°S, 28°N, and 44°N latitude. (See **Reading coordinates** at left.) Close each profile window when you are finished viewing it.

7. According to the map, where else do you see spreading ridges besides the Atlantic Ocean? (Turn on the **Labels - Oceans** theme if you need help identifying the oceans.)

Explaining the origin of these ocean floor features is a key element in the current understanding of Earth's structure. You will find out more about what causes spreading ridges in a later activity.

The lowest lows

Earth's lowest spots aren't in the *middles* of the oceans, so they must be elsewhere in the ocean basins.

▶ Using the Hot Link tool ⚡, click on the profile line that crosses the west coast of South America to see the topography of this area.

8. Sketch the topographic profile, and label the deepest point with its estimated depth.

These deep linear features, the lowest points on Earth, are called *trenches*.

9. About how far from the edge of the continent is the deepest part of the trench? (Estimate the distance from the topographic profile.)

▶ Close the topographic profile window.
▶ Use the elevation scale and the Hot Link tool ⚡ to examine other profiles and identify two other locations where you find trenches similar to the one near South America.

10. From the profiles, give the name and estimated depth of each trench.
    a. name =   maximum depth =
    b. name =   maximum depth =

The lowest point on Earth is called the Challenger Deep, part of the Mariana Trench in the Philippine Sea. It reaches a depth of over 11 km below sea level. The highest point on Earth is Mt. Everest in Nepal, nearly 9 km above sea level.

Do you get the picture?

Now it's time to see how well you can interpret Earth's topography. A profile line is shown on the map below.

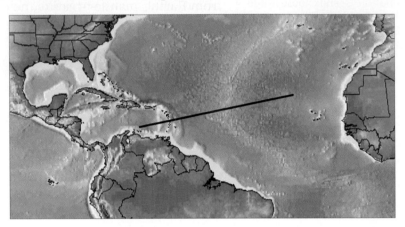

11. Without looking at the profiles on your map, sketch what you think the profile along the line above would look like. Label any ridges and trenches on your profile.

Volcanic clues

The discovery of spreading ridges in the 1940's made scientists wonder what caused these long features to form underwater in the middle of the ocean. You will investigate these next.

▶ Turn off all themes except **States**, **Countries**, and **Topography**.

▶ Turn on the **Ridge Volcanics** theme.

Ridge volcanics

Ridge flyby animation

To view a simulated "flight" along a spreading ridge, click the Media Viewer button 🎬 and choose **Ridge Flyby** from the media list.

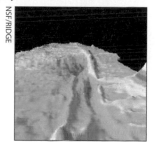

Since discovering spreading ridges in 1953, scientists have studied them extensively using radar and underwater research vessels. They have learned that spreading ridges form where rising magma—molten rock from Earth's mantle—breaks the ocean floor along long cracks, or *fissures*. When a fissure opens, magma squeezes up through it and solidifies, adding to the rock on both sides of the ridge and filling in the crack. Generally, the magma doesn't create individual volcanoes or lava flows at the ridges and the relatively gentle underwater eruptions go unnoticed. The **Ridge Volcanics** theme shows the locations of the fissures, which run down the center of the ridges.

▶ To see an animation of a ridge volcanic eruption, click the Media Viewer button 🎬 and choose **Spreading Ridge**. Close the Quick-Time Player window when you have finished viewing the animation.

To better understand ridge volcanics, look at a block diagram of the Mid-Atlantic Ridge. Note the symmetrical appearance of the ridge.

▶ Turn on and activate the **Block Diagrams** theme.

▶ Using the Hot Link tool 🔖, click on the profile line that crosses the Mid-Atlantic Ridge to view a block diagram of the ocean floor.

Features of spreading ridges

Recent research has given us a picture of the unique environments of spreading ridges. Among the unusual features of these ridges are hydrothermal vents of superheated water called "black smokers." These undersea hot springs circulate hot, mineral-rich water and dissolved volcanic gases that support an incredible array of life forms.

A spreading ridge is not "built up" layer by layer from volcanic eruptions. Rather, the ridge exists because the warmer rock near the center of the ridge is less dense than the cooler rock farther from the ridge. Buoyant forces cause this warmer rock to rise higher, forming the ridge. As fresh, hot magma is injected into the fissures, the older rock moves away from the ridge and the heat source. As it moves away, it cools, becomes denser, and sinks to form the deep ocean floor.

▶ Close the block diagram window.

Ridge volcanics aren't limited to large ocean basins.

12. Find two places where ridge volcanics lie in narrow seas between continental land masses. Circle their locations on the map below. (Hint: zoom in to take a closer look.)

A black smoker vent on the southern Juan de Fuca Ridge off the Oregon coast. To see a movie of this black smoker, click the Media Viewer button 🎬 and choose **Black Smoker** from the media list.

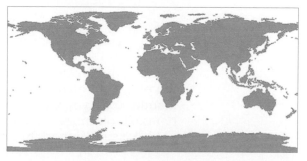

13. Based on what you have learned about ridge volcanics, are these narrow seas closing or opening wider?

Trenches and volcanic arcs

▸ Turn on the **Trenches** theme.

The **Trenches** theme shows the locations of the deep ocean trenches you learned about earlier in this activity. These trenches are the deepest parts of the ocean basins.

14. Are trenches found closer to land or closer to the mid-ocean?

▸ Turn on the **Volcanoes** theme.

In the **Volcanoes** theme, each brown triangle represents an individual volcanic feature that has been active within the past 10,000 years.

15. Ridge volcanics typically occur in the ocean. Where are the majority of individual volcanoes found?

▸ Turn off the **Ridge Volcanics** theme.

There are two fairly distinct populations of volcanoes around the world— those in clusters and those that are fairly isolated from each other.

16. In general, where are the clusters of individual volcanoes located relative to the trenches and to the edges of the continents?

Look closely at your map to find several of these clusters of volcanoes.

17. On the map below, circle the locations of four volcano clusters.

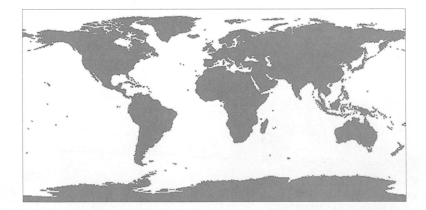

Many of the isolated volcanoes are created by "hot spots." These are hot plumes of magma that rise toward the surface from deep within the mantle. Later you will learn about the processes that create these two types of volcanoes and about differences in their characteristics. For now, it is only important to know the two groups exist.

To activate a theme, click on its name in the Table of Contents.

"I know there's a volcano near my community, but it's not on the map!"

Remember, this map only shows the volcanoes that are known to have been active in the past 10,000 years.

"Nothing happens when I click with the Identify tool!"

If the Identify Results window does not open when you click on the volcano symbol, check these things:

• Is the Identify tool ⓘ selected?
• Is the Volcanoes theme active?
• Are you clicking directly on the volcano smbol?

Volcanoes and your community (optional)

Up to now, you have been looking at the "big picture" of volcanoes. Now you will learn about the volcanic feature nearest to your community.

▶ Choose **View** ▶ **Full Extent** to view the whole map, then use the Zoom In tool 🔍 to zoom in on the area in which you live.

▶ Activate the **Volcanoes** theme.

▶ Using the Measure tool 📐, click first on your community and move the cursor to the volcanic features nearest to your community. Read the distance in the status bar. When you have found the nearest feature, double-click on it.

18. How far, in kilometers, is your community from the nearest volcanic feature?

▶ Using the Identify tool ⓘ, click on the volcanic feature you identified as the nearest to your community to learn more about it.

19. Describe the name (if any), location, date of last eruption, and any other information you learn about the volcanic feature nearest your community.

Seismic clues

Previously, you looked for patterns in topography and the locations of volcanic features to understand Earth's changing surface. Now you will look for patterns of earthquakes around the world and explore their relationship to topography and volcanoes.

▶ Turn off all themes except **States**, **Countries**, and **Topography**.
▶ Turn on the **Latitude** / **Longitude** theme.

What is an earthquake?

An earthquake is the sudden breaking of rock that results from the slow accumulation of stress in the rocks. However, only brittle, cold rocks can be broken in earthquakes. If rock is heated, it can stretch and deform slowly over time without breaking. The size or magnitude of an earthquake is determined by how much rock is broken in an event.

Earthquake depth

The data in this theme are displayed according to depth. The same data are displayed according to earthquake magnitude in the **Earthquake Magnitude** theme, which you will explore later.

▶ Turn on and activate the **Earthquake Depth** theme.

This theme shows the locations of nearly 8,000 earthquakes recorded around the world since 1973. The data include all magnitude 5 and larger earthquakes from 1973-2000, plus a few smaller events from regions that don't have large earthquakes.

Viewing the map as a globe

There are disadvantages to viewing the Earth as a flat map. It's not obvious that a feature that disappears off the left edge of the map continues on the right edge, or that features at the top or bottom actually meet at the poles. The real Earth is nearly a sphere, and the earthquake patterns you are looking for may be easier to see if you view the earth as a globe. This view is called an **orthographic projection**.

▶ To view the earthquakes on a globe:

- Click the Orthographic Projection button 🔘 to view the world as a sphere.
- Click the rotation buttons ⬅️ ⬆️ ⬇️ ➡️ to turn the globe 10° per click in any direction. For example, click the ⬆️ or ⬇️ button nine times (9 x 10° = 90°) to view Earth from the poles.

▶ In addition to changing the projection, you can use the Zoom In tool 🔍 and the Pan tool ✋ to take a closer look at the distribution and patterns of earthquakes.

20. Describe any patterns you see in the distribution of earthquakes over the Earth's surface. (For example, do they form lines, arcs, circles, or clusters? Are the patterns connected or disconnected?)

Disappearing topography

Don't panic when the topography disappears when you switch the view to Orthographic projection. The Topography is an image theme, and can't be reprojected. It will reappear when you switch back to geographic projection.

Earthquake depth patterns

Next, you will look for patterns in the depth of earthquakes around the world.

▶ Examine the **Earthquake Depth** legend in the Table of Contents.

21. What color are the *shallowest* earthquakes? What color are the *deepest* earthquakes?

    a. shallowest –

    b. deepest –

▶ Click the Geographic Projection button 🖥️ to return to a rectangular projection.

▶ Turn on the **Trenches** and **Ridge Volcanics** themes.

▶ Zoom in on South America and look at the ridge and trench that bound the continent.

Look closely at the ridges and trenches—the earthquake depth patterns at these two types of features are different.

22. How deep are the earthquakes near each type of feature? (Give a range, from the shallowest to the deepest.)

    a. ridges

    b. trenches

Next, you will look at differences in earthquake depth around the world.

▶ Use the Zoom In tool 🔍 and the Pan tool 🖐 to look closely for regions with deep earthquakes.

23. Are there any earthquakes near spreading ridges with depths greater than 150 km? Why do you think this is?

Where am I?

If you need help with the names of countries for question 24, turn on and activate the **Countries** theme and use the Identify tool 🛈 to click on a country to read its name.

24. Identify three areas with many earthquakes deeper than 150 km. (For example, "The west coast of Central America near Mexico, El Salvador, and Nicaragua.")

▶ Zoom in closely on the areas you identified in the previous question to examine the pattern of the earthquakes.

25. In general, how do the earthquake depths change as you move farther away from the trench?

▶ Turn on the **Volcanoes** and **Trenches** themes.
▶ Click the Zoom to Full Extent button 🔳 to view the whole map.
▶ Turn on and activate the **Block Diagrams** theme.
▶ Using the Hot Link tool ⚡, click on each of the profile lines that crosses a trench to view a block diagram at that location.

26. What do you think is happening at the trenches to cause the deep earthquakes there? Why don't many deep earthquakes occur elsewhere?

▶ Turn off the **Earthquake Depth, Latitude / Longitude, Block Diagrams**, and **Volcanoes** themes.
▶ Turn on the **Ridge Volcanics** theme.

Earthquake magnitude patterns

Next, you will look for patterns in earthquake magnitudes around the world.

▶ Turn on and activate the **Earthquake Magnitude** theme.

▶ Examine the **Earthquake Magnitude** legend in the Table of Contents.

▶ Use the Zoom In tool 🔍 and the Pan tool 🖐 to look more closely for regions with large earthquakes.

27. Where do you see more large earthquakes (Magnitude 7 or higher)—near the spreading ridges or near the deep trenches?

Comparing depth and magnitude

To see if there a relationship between the magnitude and depth of earthquakes, you will hide all of the earthquakes except the largest and deepest and look for a connection between them.

▶ To show only the earthquakes of magnitude 7 or higher, you will use a theme property query operation.

• Activate the **Earthquake Magnitude** theme.

• Choose **Theme ▶ Properties** and click the query builder button 🔨.

• Enter the query ([Magnitude] >= 7.0) as shown below and click **OK** twice to display only the selected earthquakes.

What's a query?

A query is a question. In this case, it's a question that you ask about the features on a map. Usually, you want to know which map features meet certain requirements. In this case, the question is "Which earthquakes had a magnitude of 7 or higher?"

When you ask a computer a question, you must phrase it in a way that the computer understands. The Query Builder helps you phrase your question correctly, but it's still easy to make a mistake and get a "syntax error". If you make a mistake, don't panic—just start over and re-enter your query statement. Pay attention to spelling and punctuation, and especially to parentheses and brackets.

ArcView allows two kinds of queries. A normal query shows all of the features and highlights the ones that meet your requirements. A theme properties query like the one used here is similar, but it hides all of the features that *don't* meet your requirements. It's most useful when there are lots of features or the features are very close together.

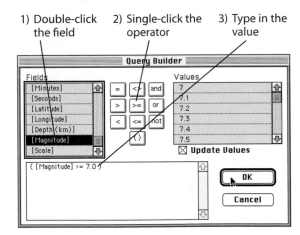

▶ To display only the earthquakes that occurred at depths of 150 km or more, use another theme property query:

• Turn on and activate the **Earthquake Depth** theme.

• Choose **Theme ▶ Properties** and click the query builder 🔨 button.

• Enter the query ([Depth (km)] >= 150) and click **OK** (twice) to display only the selected earthquakes.

▶ Turn the **Earthquake Depth** and **Earthquake Magnitude** themes on and off to explore their relationship.

28. Compare the locations of large earthquakes (magnitude 7 or greater) to the locations of deep earthquakes (depths greater than 150 km). Do they occur together, or are they found in different regions? Explain.

29. Spreading ridges were formed recently from hot magma. How might this affect the magnitude and depth of earthquakes near the ridges?

30. Mountain ranges, trenches, and ocean ridges all have large variations in topography. How are these features related to the distribution of earthquakes, volcanoes, and volcanic ridges?

Activity 1.3

Earth's inner structure

In addition to differences in composition, Earth's inner structure is defined by differences in the state of the material. The interior of Earth is still hot enough for the outer core to be molten and for the mantle to flow like putty.

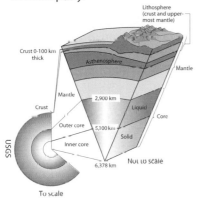

Discovering plate tectonics

The theory of **plate tectonics** states that the Earth's outermost layer is fragmented into a dozen or more large and small plates that are moving relative to one another. Plates, also called **lithosphere**, are rigid layers made up of the crust and uppermost rigid part of the mantle. The plates ride on top of the hotter, more mobile material in the deeper mantle.

Heat and gravity are the two main driving forces of plate tectonics. Earth's core reaches temperatures over 6,000°C. About half of this heat comes from the decay of naturally occurring radioactive minerals. The other half is left over from the heat generated when Earth formed. Heat energy travels slowly from the interior to the surface by the processes of convection and conduction.

In conduction, heat energy moves by collisions between atoms. Convection occurs as rock heats up, expands and becomes less dense. Gravity forces this buoyant rock to rise toward the surface. There it cools, becomes denser, and eventually sinks deep into the mantle again. Convection cools Earth more efficiently than conduction. Still, it takes a long time to cool an object the size of Earth.

At mid-ocean ridges, hot rock rising from deep in the Earth forms new oceanic lithosphere. This lithosphere stands about 2 km higher than the average elevation of the ocean basins. Gravitational forces push the plates down and away from this high ridge. At trenches, old, cold and therefore dense lithosphere is diving into the deeper mantle. Gravity forces pull the dense lithosphere into the mantle. The resulting ridge-push and trench-pull forces created by heat and gravity are driving plate tectonics.

Evidence of change

Earthquakes - reshaping the land

Earthquakes occur when stresses build up in cool, brittle rock and are released suddenly as the rock breaks. These stresses are often found at the edges of plates, where the plates move together, pull apart, or grind past one another.

A major factor controlling where earthquakes occur and the amount of energy released when they fail is rock strength. Rock strength depends on the temperature and the rate at which the rock is deformed. Rocks near the surface are cool and brittle. When stressed beyond their mechanical strength, they break like a rigid pencil, sending out shock waves in all directions. Typically, rocks below 20–30 km depth are warmer and more ductile than the shallower rocks. When they are stressed, they deform rather than break.

Earthquakes often cause changes in topography. Over millions of years, they can produce great mountain ranges and large valleys.

A globe-circling "wound"

American oceanographer Bruce C. Heezen described the mid-ocean ridge system as "the wound that never heals." Hidden beneath the ocean surface, these ridges extend for thousands of kilometers forming an interconnected network that encircles Earth.

Hotspot volcanism

A subduction zone

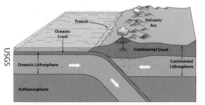

A subduction zone occurs at a convergent plate boundary, where a denser lithospheric plate plunges beneath a less dense lithospheric plate.

Volcanoes - creating new crust

Volcanoes are another indicator of changes occurring in Earth's crust due to plate tectonics. Volcanoes develop where molten rock rises from deep inside Earth to the surface, then cools and forms new lithosphere.

Most of the world's volcanic activity occurs along the spreading ridges found throughout the world's ocean basins. At spreading ridges, hot rock rises to form new oceanic lithosphere. As the magma rises beneath the ridge, the plates pull apart at the fissure running down the center of the ridge. Magma squeezes into the fissure and cools, solidifying onto the edges of the two plates. Gravity pulls the plates slowly down and away from the ridge's center in a conveyor belt-style process.

Pulses of volcanic activity along the ridges are separated by quiet intervals with no activity. On human time scales, the process appears irregular, but on geologic time scales of millions of years, it is steady.

Hot spot volcanoes

Another type of volcano forms above hotspots, hot plumes of magma that rise from deep within Earth's mantle and burn through the crust. As plates move over the stationary hotspots, the magma plume continues to burn through the crust, creating chains of volcanoes. The ocean floor has many chains of hotspot volcanoes. Though well known, hotspots like Hawaii and Iceland are the least common type of volcano.

Subduction zones - recycling crust

Gravity simply won't allow Earth to expand like a balloon. Thus, when new lithosphere is created at spreading ridges, it must either be compressed or recycled somewhere else. Which process occurs depends on the lithosphere's density. High density oceanic lithosphere is recycled, while low density continental lithosphere is compressed.

Oceanic lithosphere is recycled into the mantle in a process called **subduction**. At subduction zones, gravity pulls the denser oceanic lithosphere deep into the mantle, forming trenches of the ocean floor up to 11 km deep. Because of density differences, trenches often develop near the boundary between oceanic and continental lithosphere.

As oceanic lithosphere is subducted, it is warmed by the surrounding mantle rock. At a certain depth, the pressure causes water trapped in the minerals of the subducting plate to be released, lowering the surrounding rock's melting point and causing it to melt. The molten rock rises, burns through the overriding plate, and creates a chain of volcanoes on the surface called a volcanic arc.

Continental mountain ranges

When two continental plates collide, the low-density rocks are too buoyant to be subducted. Instead, they smash into one another, forming huge mountain ranges. For example, the Himalayas are being thrust upward by the collision between the Indian and the Asian plates. The Appalachian mountains of the Eastern US are an eroded remnant of a similar collision in the past. When first formed, they may have rivaled the Himalayas in size.

Classifying plate boundaries

Using earthquake, volcano and topographic patterns around the world, scientists mapped the major plate boundaries. Once the boundaries were identified, it was time to figure out how the plates interact and use this information to classify each boundary. There are three basic types of plate boundaries, identified according to the plate motions at the point of contact.

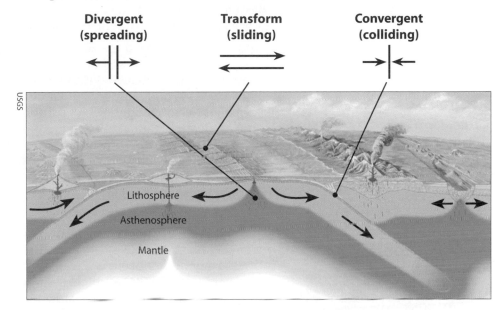

Divergent boundaries

At divergent boundaries (such as mid-ocean spreading ridges), plates are moving away from each other. The plates in these areas are relatively thin (less than about 30 km), weak, and warm; earthquakes here are generally shallow and small to moderate in size. As the plates split, the gap is filled with molten rock that forms a high ridge. The age of the rock bordering divergent boundaries increases away from the boundary. Occasionally, divergent boundaries develop within continental plates, causing them to split apart. This is happening in East Africa today, and may eventually create a new ocean basin and spreading ridge there.

Convergent boundaries

At convergent plate boundaries, plates collide and deform. The most common type of convergent boundary is a subduction zone. A subduction zone is characterized by a strong, cold oceanic plate diving into the mantle and forming a deep trench along the entire boundary.

Continental plates are very buoyant and cannot be subducted. When continental and oceanic plates collide, the oceanic plate will always be subducted because it is more dense (>3.0 g/cm³ vs. 2.7). Where two continents collide, the plates crumple and build large mountain ranges like the Himalayas. Oceanic plate gets older and becomes colder and more dense as it moves away from the spreading ridge. Thus, if two oceanic plates collide, the older, denser plate will subduct into the mantle.

Types of plate collisions

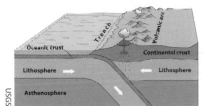

Oceanic–Continental collision

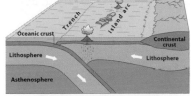

Oceanic–Oceanic collision

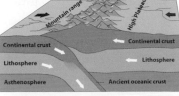

Continental–Continental collision

Volcanoes are created along the plate boundary where the oceanic plate descends into the mantle, carrying with it minerals that have water bound in their crystal structure. As the descending plate reaches a depth near 150 km, these "hydrated" minerals release the bound water, which mixes with the mantle rock, allowing it to melt more easily. The lower-density molten rock melts its way upward from the mantle into the overriding plate, forming volcanoes where it reaches the surface. These volcanoes pose a hazard to humans because they're often very explosive and located near population centers. Mt. St. Helens and Mt. Fuji are examples of this type of volcano.

Earthquakes occur along the zone of contact between the two converging plates and at great depths in the subducting slab. Rock conducts heat very poorly, so the subducted slab stays cold and rigid for hundreds of millions of years, and hundreds of kilometers deep. At subduction zones, earthquakes occur along the cold, brittle subducting plate. This pattern of shallow to deep earthquakes is an identifying characteristic of subduction zones. Most of the world's largest earthquakes occur along subduction zones, where the contact area of rock against rock is both deep and wide.

Transform boundaries

Transform boundaries occur where the edges of two plates grind past each other without converging or diverging. They are characterized by a strong plate capable of breaking in large earthquakes. The San Andreas Fault in California is a transform plate boundary. The linear features in the seafloor that offset segments of the spreading ridges are also transform faults. At transform boundaries, the thickness of the lithosphere ranges from a few kilometers near ocean ridges to 100 - 150 km where they cut across continents.

Earthquakes occurring at transform boundaries in the seafloor can be very large if the rupture occurs along a great length, but are usually moderate and of shallow depth (less than 30 km). The weaknesses created by the faulting may help provide routes for magma to reach the surface more easily, but volcanoes along these boundaries are not common.

Questions

1. What two forces drive the processes that change Earth's surface?

2. Knowing that the lithosphere "grows" at spreading ridges, why would we suspect that trenches and subduction zones exist, even if we can't see them?

3. Why is the center of a spreading ridge higher than the surrounding ocean basin?

4. At what types of plate boundary would you find a narrow band of shallow earthquakes?

5. At what type of boundary would you find a wide band of earthquakes, extending to depths greater than 150 km?

6. Why are most subduction zones located along the edges of continents?

How does water help rock melt?

Although it seems against common sense, mixing two substances often lowers the melting point of one or both substances. This is the case with rock, which melts at a lower temperature when water molecules are added to the mix.

A transform boundary

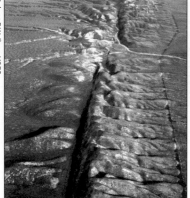

Robert E Wallace, USGS

The San Andreas fault, a continental transform boundary, is visible in this aerial photo near San Luis Obispo, California.

Other evidence for plate tectonics

In Unit 2, you will investigate magnetic patterns in oceanic rocks that provide evidence of plate tectonics. Other evidence that you won't explore in this activity include paleoclimate and fossil evidence.

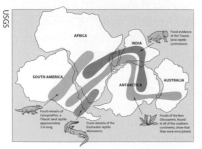

USGS

Fossil evidence shows that present-day continents were once part of a larger super-continent, now called Pangaea.

Activity 1.4

Analyzing plate boundaries

You have looked at patterns in earthquakes, volcanoes, and global topography and learned how these are related to plate movements. Using all of this evidence, predict where you think the surface of Earth is deforming along plate boundaries.

▶ Launch the ArcView GIS application, then locate and open the **dynamic.apr** project file.

▶ From the list of views, open the **Clues** view.

Remember, as you work through this activity, you can change your view of Earth at any time from a flat map to a globe view (and back again).

▶ To view the Earth as a globe:

• Click the Orthographic Projection button [O] to view the world as a sphere.

• Click the rotation buttons [←][↑][↓][→] to rotate the globe 10° per click in any direction.

• To return to a rectangular map, click the Geographic Projection button [□].

Locating plate boundaries

▶ Turn themes on or off to help you locate the plate boundaries.

1. Look at all the available evidence and use it to determine where you think Earth's plate boundaries are located. Draw the boundaries on the map on page 25. Use a solid line where you're sure of a boundary and a dashed line where you're not sure.

2. Add a legend to your map to show what your symbols mean.

Comparing to "accepted" plate boundaries

After you have drawn your map of plate boundaries, compare it to the currently accepted plate boundary map.

▶ Close the **Clues** view and open the **Locating Boundaries** view.

▶ Turn on the **Plate Boundaries** theme in the table of contents.

3. On your map, circle the areas where your boundaries are different from the accepted boundaries.

4. What do you think accounts for the differences between the plate boundaries you drew and the accepted plate boundaries?

To turn a theme on or off, click its checkbox in the Table of Contents.

To open a view, select the name of the view in the project window and click the Open button.

There is disagreement within the scientific community on the location and existence of several plate boundaries. This is especially true in areas at the far northern and southern latitudes, where few instruments are available to monitor earthquakes and plate motion, and in the Indian Ocean and the Mediterranean. In those two areas plates are moving slowly, and it's especially difficult to accurately measure plate motion.

To turn a theme on or off, click its checkbox in the Table of Contents.

▶ Turn on the **Earthquakes** theme.

5. Plate Tectonics theory says that plates are rigid and will only deform at their boundaries. Earthquakes are indicators that a plate is deforming. How well does Plate Tectonic Theory predict the location of earthquakes? Explain.

Identifying plate boundaries

Using earthquake data, magnetic patterns in the seafloor, and precise satellite navigation systems, researchers have collected information about how the plates are moving relative to each other.

▶ Turn off the **Earthquakes** theme.

▶ Turn on and activate the **Relative Motion** theme.

To activate a theme, click on its name in the Table of Contents.

The arrows in this theme are called *vectors* and are shown in pairs on opposite sides of the plate boundaries. They indicate the direction and speed of each plate relative to the other across the boundary. In each pair, the length of the arrows (the rate of plate motion) is the same, but their directions are opposite.

"Where do I click?"

When using the Identify tool on the Relative Motion theme's arrows, clicking on the arrow symbol can be somewhat tricky. You need to click on the center of the arrow. Because the "arrowhead" is longer than the "shaft," the center is not quite where you might expect it to be. The diagram below shows where to click.

▶ Zoom in 🔍 and use the Identify tool 🔘 to click on an arrow symbol to get precise rate information for that location. Close the information window when you are finished with it.

Using these vectors, you can determine if the two plates meeting at a boundary are colliding (convergent boundary), moving apart (divergent boundary), or slipping past each other (transform boundary). You will also need to recall what you've learned about the seismic, volcanic, and topographic patterns related to each type of boundary. These may help you identify plate boundaries where motion vectors are absent or uncertain.

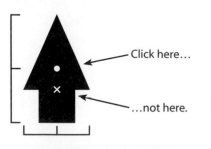

6. On the map you created in question 1, identify the type of each boundary as convergent, divergent, or transform using a symbol or color key. If you are not sure which type a particular boundary is, use your best guess.

7. Add the boundary type information to your map legend.

Hot tip for Hot Links

When you use the Hot Link tool ⚡:

• The theme containing the hot links must be active (the tool on the tool bar will be dark ⚡, not grayed out ⚡.)

• Click when *the* tip of the lightning bolt cursor is over the feature.

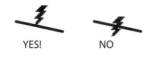

YES!   NO

Refresh your memory!

Question 10 may be easier to answer if you go back and look at the features in the **Clues** view.

▶ See how well you predicted the types of boundaries.

• Close the **Locating Boundaries** view window and open the **Boundary Types** view.

• Turn on and activate the **Plate Boundaries** theme.

• To see a diagram of each boundary type, click on a plate boundary using the Hot Link tool ⚡.

• Close windows when finished.

8. Mark the areas where your boundaries differ from the accepted map. Estimate the percentage of boundaries that you identified correctly and record it here.

9. How do you account for the differences between your predictions and the actual boundary types?

▶ Select a plate boundary of your choice within the view, and zoom in on it.

▶ On your paper map, draw a colored box around the plate boundary you have chosen.

10. Using all the data available in this activity, list and describe the evidence that indicates the type of boundary you have chosen. (Hint - don't forget to include specific seismic, volcanic, and topographic clues.)

Plate Boundaries

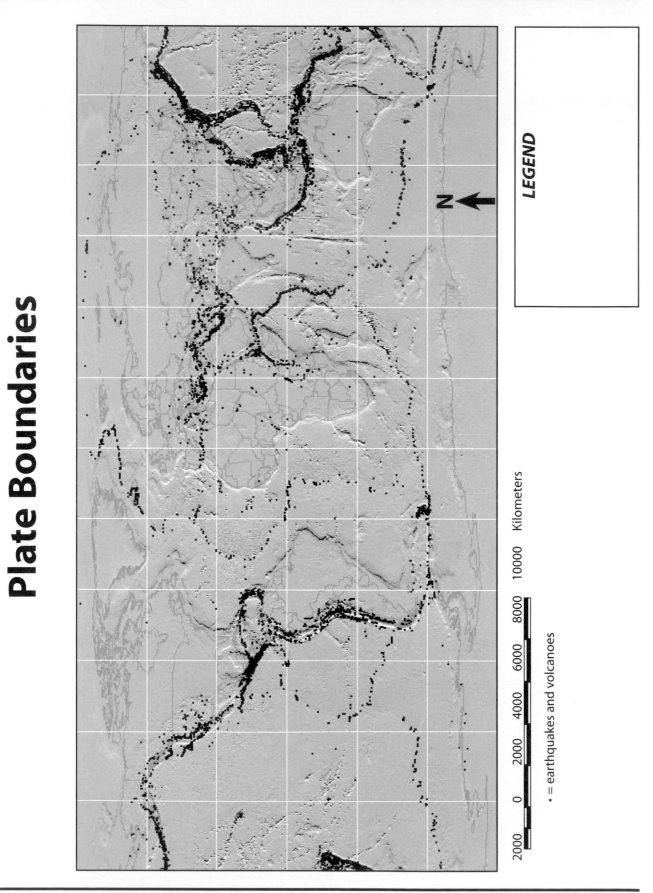

LEGEND

2000   0   2000   4000   6000   8000   10000   Kilometers

• = earthquakes and volcanoes

N

Unit 2
Exploring Plate Tectonics

In this unit, you will...

- *calculate the rate of spreading of the Atlantic Ocean,*

- *investigate whether plate spreading rates change with time or vary across the globe,*

- *predict the ultimate fate of the Juan de Fuca plate,*

- *use the Hawaiian islands to determine the rate of motion of the Pacific tectonic plate, and*

- *determine when San Francisco and Los Angeles will finally be neighbors across the San Andreas Fault.*

Aerial view of the San Andreas fault slicing through the Carrizo Plain east of the city of San Luis Obispo, California.

Activity 2.1   # Testing plate tectonics

You now know that Earth's plates are constantly growing, shrinking, and moving around. This means that the Earth must have looked very different in the past.

Questions for thought and discussion

1. What evidence would you look for to show that the continents were once in very different locations than they are today? List as many signs of these changes as you can.

Answer the following questions on your own or in a small discussion group, as assigned by your instructor.

2. Where would you go to find the evidence described above, and how would you proceed to look for it?

Thinking deeper (optional question)

Some scientists have suggested that even small changes in the shape, location, orientation, or topography of continents or oceans could significantly affect life on Earth.

3. Describe how even a small change in the Earth's surface might have a major effect on life on the planet. Draw pictures to illustrate your point.

Activity 2.2

Investigating seafloor age

In this activity, you will investigate evidence that tectonic plates change in size, shape, and location over time. This evidence suggests that plate tectonics has occurred for hundreds of millions of years.

▶ Launch the ArcView GIS application, then locate and open the **dynamic.apr** project file.

▶ From the list of views, open the **Changing plates** view.

This view shows differences in the age of the rocks that form the ocean floor. These ages were determined using data from several sources, including measurements of weak magnetic patterns in the minerals of the rocks and determining age dates from drill cores. These age data have been summarized, mapped, and color-coded. The age bands generally run parallel to the spreading ridges.

Spend a few minutes examining the seafloor age data. Use the **Seafloor Age** legend to answer these questions.

1 How many years does each colored band represent?

2. What happens to the age of the seafloor beneath the Atlantic Ocean as you get farther from the mid-Atlantic ridge?

3. How long ago did the North Atlantic Ocean begin to open up near what is now the East Coast of the United States? Describe the reasoning you used to arrive at this answer.

4. Did the North Atlantic Ocean basin open up at the same time as the South Atlantic Ocean basin near South America? Describe evidence that supports your answer.

5. How could you use the widths of the age bands to figure out if the rate of plate spreading has been constant over time?

Viewing Earth as a globe

As you work through this activity, you can change your view of the Earth at any time.

To view the world as a globe:
Click the Orthographic Projection button ▢.

To rotate the view:
Click the rotation buttons. Each click rotates the view 10 degrees in that direction.

←↑↓→

To view the world as a rectangle:
Click the Geographic Projection button ▢.

6. The oldest continental rocks formed 3.8 billion years ago. What is the age of the oldest seafloor?

7. Why do you think there is such a difference in age between the oldest seafloor and the oldest continental lithosphere? (Hint: Near which type of plate boundary do you find the oldest oceanic rocks?)

8. During the time since the oldest continental rocks formed, about how many times could the world's ocean floors have been completely "recycled"? How did you determine this number?

Sea floor age is a critical piece of evidence for plate tectonics. It shows us where new sea floor is created and how often. It can also be used to reconstruct how ocean basins have grown in the past, and how they may change in the future.

Activity 2.3   # Determining seafloor age

Paleomagnetism

For reasons that are not fully understood, Earth's magnetic field changes polarity. That is, the north and south magnetic poles reverse. This process has occurred at uneven intervals, roughly once every 250,000 years over the past 5 million years.

As the ocean floor spreads at mid-ocean ridges, magma from the mantle solidifies onto the plate near the ridge. This process forms volcanic rocks in the oceanic lithosphere.

When the newly-formed lithosphere cools, some of the minerals become magnetized and align with Earth's magnetic field. This "fossil magnetism" preserves a record of the direction of the Earth's magnetic field at the time the rocks formed.

Some bands of oceanic lithosphere formed when Earth's magnetic field was like it is today. We call this "normal polarity." Other bands of oceanic lithosphere formed when the Earth's magnetic field had the opposite polarity. We say these bands have "reversed polarity."

The polarity changes of Earth's magnetic field are not regular. Instead they make complex sequences of normal and reversed polarity. The combination of seafloor spreading and changes in magnetic polarity form a unique pattern similar to the bar codes on products you buy at the store.

Ships carrying sensitive equipment have crisscrossed the oceans recording the patterns of these fossil magnetic fields. Geologists familiar with the polarity time scale can read the code and use it to determine the age of the oceanic lithosphere.

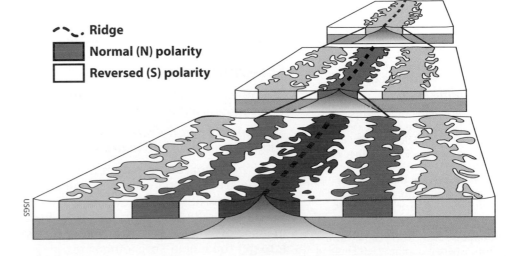

- ⌒⌒. **Ridge**
- ▨ **Normal (N) polarity**
- ☐ **Reversed (S) polarity**

USGS

Constructing continents

Some plates remain relatively unchanged for millions of years, but most are growing or shrinking in size. Over Earth's 4.8 billion-year history, many oceanic plates have been destroyed and new ones have been created. Continental plates have been torn apart and smashed together in new arrangements, as in the Himalayas. Plate tectonics is reshaping the face of our planet.

Relative motion

All of Earth's plates are in motion. For this reason, it's often easiest to describe a plate's motion relative to a neighboring plate.

For example, if you were standing on the eastern edge of the African plate looking at your dog Ralph (sitting very patiently) over on the bordering Indo-Australian plate, Ralph would appear to be moving away from you. To Ralph, he is the one sitting still while you move away from him. The motion is relative to the observer's point of view, or *frame of reference*.

Relative plate motion

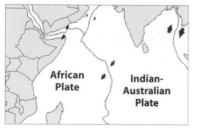

Arrows show the motion of the African plate and the Indian-Australian plate relative to each other.

Absolute motion and hot spots

When talking about entire plates, scientists often prefer to describe the plate motions relative to a common frame or point of reference. Ideally, this reference point is not moving like the plates but is instead stationary ("fixed") deep within the Earth.

There are twenty or so places around the Earth where the mantle below the lithosphere is unusually hot. These "hotspots" partially melt the plates above them, creating topographic bulges and chains of volcanoes. While their origin is not completely understood, these hotspots appear to be fixed in the deep Earth and provide a frame of reference for measuring "absolute" plate motions.

Two of these hotspots are located in the United States. One is at the island of Hawaii at the southeast end of the Hawaiian Island chain. The other is located at Yellowstone National Park and is responsible for the volcanic and geothermal activity in that region.

Studying plate motions in a fixed frame of reference provides scientists with a better understanding of the forces that drive plate tectonics.

Absolute plate motion

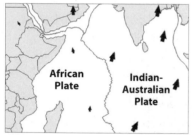

Arrows show the motion of the African plate and the Indian-Australian plate relative to an absolute frame of reference (a hot spot to the south).

Spreading rates

Scientists measure spreading rates of oceanic lithosphere near spreading ridges as either a "half" or a "whole" rate.

The **half-spreading rate** is the rate at which new oceanic lithosphere is added to either of two plates separated by a spreading ocean ridge.

The **whole-spreading rate** is the rate at which two plates separated by a ridge are moving away from one another. Usually, equal amounts of new oceanic lithosphere are added to each plate as they move apart. The whole-spreading rate is therefore twice the half-spreading rate. This means you can measure the half-spreading rate for one plate and double it to get the whole-spreading rate.

A Titanic example of spreading rates

The wreck of the Titanic rests at the bottom of the Atlantic Ocean to the west of the Mid-Atlantic Ridge. It moves about 15 mm farther west of that ridge each year. So the half-spreading rate at this location is 15 mm/yr. Thus, the whole-spreading rate is 2 x 15 mm or 30 mm/year.

So, as the Atlantic Ocean widens, the Titanic is slowly being carried away from where it originally left England at a rate of about 3 cm (a little over an inch) per year!

Disappearing islands

The Hawaiian and Emperor Volcanic chains mark the motion of the Pacific plate over a stationary hotspot. As each volcanic island is carried past the hotspot, wave action quickly erodes the exposed island completely away, leaving a submerged seamount.

Changes in plate motion

Spreading rates of spreading ridges and the directions of plate motions have changed many times during the past 200 million years. The Pacific Plate experienced a particularly dramatic change in direction about 43 million years ago. If you examine the pattern of volcanoes extending northwest from Hawaii in the map below, you will notice that they form two intersecting chains, the Hawaiian and the Emperor Volcanic Chains.

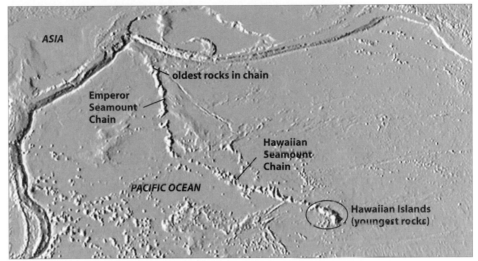

Seafloor topography of the northern Pacific basin

By measuring the ages of rocks in each chain, we know that the "bend" in the chain occurred around 43 million years ago. Prior to that time, the Pacific Plate was moving in a north-northwest direction, then abruptly (in geologic time, anyway) changed to a more west-northwest direction. Other seamount chains in the Pacific show a similar pattern.

Earth's heat engine—slowly "running out of gas"

As Earth cools, the convection that drives the plates is gradually slowing down. Eventually—in billions of years—Earth will have lost most of its internal heat to space, and plate motions will slowly grind to a stop.

Sudden changes in plate motion occur when two or more continents collide. Continental plates are too buoyant to be subducted. Instead, they plow into each other and pile up to form mountains. The Himalaya mountains, the highest on Earth, are being thrust upward as the Indian and the Asian plates collide. This has been going on continuously for the past 50 million years. Since the two plates first made contact, the rate of sea floor spreading in the Indian Ocean has decreased considerably.

Questions

1. How are magnetic minerals in the oceanic lithosphere used to determine its age?

2. What is the difference between the relative motion and the absolute motion of a plate?

3. If the half-spreading rate at a plate boundary is 3 cm/yr, what would the whole-spreading rate be at that boundary?

4. What might cause a plate to slow its movement or change direction?

Activity 2.4

Investigating plate motion

In this section, you will measure the rate at which new lithosphere has formed along the Mid-Atlantic Ridge over the past 140 million years. Then, you will examine how this rate changed as the Atlantic Ocean opened.

Rate of spreading

To calculate the whole-spreading rate, you need to know the total width of new lithosphere formed and the time it took to create it.

my = million years

$$\text{whole-spreading rate} = \frac{\text{total width of new lithosphere (km)}}{\text{time required to create new lithosphere (my)}}$$

An easy way to do this is to measure the *half-spreading rate*—the width of new lithosphere created on one side of the spreading ridge—and double it.

Measuring new lithosphere

To turn a theme on or off, click its checkbox in the Table of Contents.

▶ Launch the ArcView GIS application, then locate and open the **dynamic.apr** project file.

▶ From the list of views, open the **Changing plates** view.

▶ To determine the width of new lithosphere created in the North Atlantic Ocean during each 20-million year time interval:

Locator map

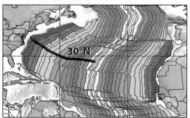

Path for measuring the spreading rate of the Mid-Atlantic Ridge in the North Atlantic Ocean.

- Turn on the **Latitude/Longitude** and **Plate Boundaries** themes.
- Using the Zoom In tool 🔍, zoom in to the North Atlantic Ocean (around 30° N latitude) between North America and Africa. (See locator map at left).
- Using the Measure tool 📐, click on one side of an interval, drag across to the other side of that interval, and double-click. For the 0–20 million year interval (lightest), measure from either edge of the interval to the Mid-Atlantic Ridge (thick line), as shown below. You should measure parallel to the transform faults, as shown in the illustration below.

Measuring the first seafloor age interval

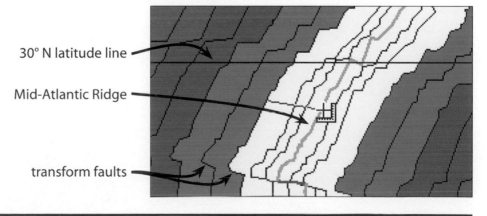

30° N latitude line

Mid-Atlantic Ridge

transform faults

▶ Read the width (Length) of new lithosphere created during that age interval on the status bar. (Your length will be different.)

Segment Length: 281.71 km Length: 281.71 km

1. Record your results in the **Width** column of the table below. Repeat this measurement for each of the other time intervals.

Time before present (my)	Width (km) measured	Half-Rate (mm/yr) calculated	Whole-Rate (mm/yr) calculated
0 - 20			
21 - 40			
41 - 60			
61 - 80			
81 - 100			
101 - 120			
121 - 140			

Computing the rates of spreading

Following the example shown at the left, perform the following calculations for each interval.

2. Calculate the half-spreading rate, in mm/yr, and enter your answer in the **Half-Rate** column. Convert units as necessary.

3. Calculate the whole-spreading rate, also in mm/yr, and enter your answers in the **Whole-Rate** column of the table.

Calculating spreading rates

Here's an example of calculating the spreading rates. Assuming the measured width is 260 km...

Width = 260 km

Half-Rate = Width/ Time

= (260 km x 1,000,000 mm/km) / 20,000,000 yrs

= 13 mm/yr

Whole-Rate = Half-Rate x 2

= 13 mm/yr x 2

= 26 mm/yr

Time before present (my)	Width (km) measured	Half-Rate (mm/yr) calculated	Whole-Rate (mm/yr) calculated
0 - 20	260	13	26
21 - 40			

4. Use a bar graph to plot the half-rate versus time before present on the chart below.

Half-Rate versus Time Before Present

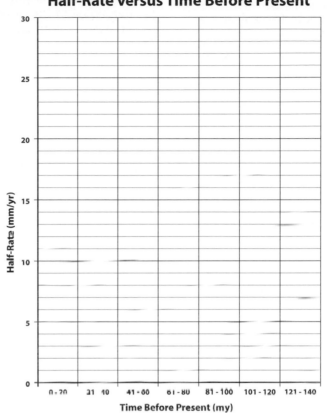

Units

Note that km/my and mm/yr are the same units!

5. How has the spreading rate changed over time?

Compare the Atlantic to the Pacific

Are all spreading ridges alike? Next, you will compare the seafloor age patterns in the southern Pacific and southern Atlantic oceans.

You could calculate the rate of spreading in the southern Pacific Ocean as you did for the South Atlantic Ocean, but you can also compare the rates by comparing the width of the age bands in the South Pacific to those in the South Atlantic. This map distorts distances at different latitudes, so compare age bands at similar latitudes.

6. Which ridge is spreading faster? How can you tell?

7. Examine the seafloor ages in the southern Pacific Ocean. Was the spreading rate constant over time? Describe any changes in the spreading rate that have occurred over the past 120 my.

Reading sea floor age

To read the age of the sea floor at any location, activate the **Sea Floor Age** theme and click on that location using the Identify tool ⊙.

Which ocean is this?

If you aren't sure which ocean is which, turn on the **Oceans** theme.

The Juan de Fuca Plate

The Juan de Fuca Plate is located off the western coast of North America.

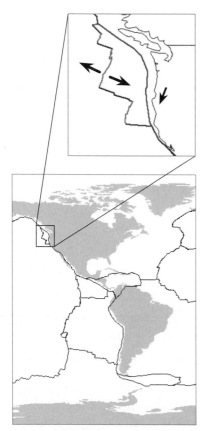

The age of the oldest rocks in an ocean basin indicates when the basin began to form.

8. Find the oldest rocks in each ocean basin. Which ocean basin has the oldest of the old rocks? (This is the oldest ocean basin.) Which has the youngest of the old rocks? (This is the youngest ocean basin.)

   a. oldest -

   b. youngest -

Juan de Fuca's disappearing plate trick

The relative youth of the oldest oceanic lithosphere (200 million years) tells us that over the course of Earth's 4.8 billion-year history, many plates have come and gone. An example of a plate destined to disappear is the Juan de Fuca plate, located off the coast of Washington and Oregon. Assuming this plate continues moving at its current velocity, it should be possible to estimate when it will completely disappear.

▶ Choose **View ▶ Full Extent** to see the entire map.

▶ Use the ⊕ tool to zoom in on the Juan de Fuca plate. It is the small plate just off the northwestern coast of the US (see map at left).

9. On the map below, label the types of boundaries and the names of the adjacent plates.

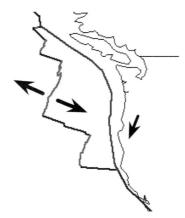

10. Under which plate is the Juan de Fuca plate being subducted?

▶ Use the Measure tool 📐 to measure the widest part of the plate in an east-west direction. (Measure perpendicular to the spreading ridge.)

11. Based on your measurement, how many kilometers of this plate must be subducted for the plate to disappear completely?

By measuring plate motions, scientists know that the new oceanic lithosphere is formed on the western side of the Juan de Fuca plate at a rate of about 25 mm/year, while on its eastern side, it is being subducted at a rate of about 30 mm/year.

12. Overall, how fast is this plate shrinking in an east-west direction per year?

A reminder
Don't forget to convert units where necessary in your calculations!

13. At this rate, how long will it take for the Juan de Fuca plate to disappear completely?

The Hawaiian-Emperor Volcanic Chain

You've seen that Earth's plates are in constant motion. How does this motion affect populated areas over time? In the next two sections, you will look at how plate motions are rearranging the United States. First you will examine the Hawaiian-Emperor Volcanic Chain to see what you can learn about plate motions from it.

▶ Turn off all themes except **States, Countries,** and **Topography**.
▶ Turn on and activate the **Hawaiian Volcanoes** theme. Leave the **Hawaiian Volcanoes (No Wrap)** theme alone for the moment.

The red symbols represent volcanoes in the Hawaiian-Emperor Volcanic Chain. Radioisotope dating of the volcanoes shows that they were not all formed at the same time. Notice that the chain extends past the western (right) edge of the map and continues on the eastern (left) edge. Take a moment to explore the volcanoes.

▶ Turn off the **Hawaiian Volcanoes** theme and turn on and activate the **Hawaiian Volcanoes (No Wrap)** theme.

In this theme, the volcanoes are grouped on one side of the map for your convenience.

▶ Choose **View ▶ Zoom to Themes** to zoom in on the volcanoes.
▶ Use the Identify tool ⬛ to find the age (in millions of years) of the volcano located at the northwestern-most (upper-left) end of the chain.

14. Record the volcano's name, age, latitude, and longitude in the table below.

To turn a theme on or off, click its checkbox in the Table of Contents.

To activate a theme, click on its name in the Table of Contents.

Name	Age (my)	Latitude	Longitude

15. Record the names, ages, and coordinates of five other volcanoes. Choose volcanoes along the entire length of the chain. Make sure to include the volcano at the southeastern-most end of the chain.

16. What trend do you observe in the ages of the volcanoes as you follow the chain from the northwest to the southeast?

Assuming that the hotspot that formed these volcanoes is not moving, and that the motion of the Pacific plate over the hotspot is responsible for the existence of the chain, you can determine the rate at which the plate is moving.

▶ Use the Identify tool [🛈] to locate the Kinmei Seamount on the map. It's near the bend in the chain.

17. How long ago was the Kinmei Seamount formed?

▶ Locate the island of Kilauea at the southeast end of Hawaii.
▶ Use the Measure tool [📏] to measure the distance between the Kinmei Seamount and Kilauea.

18. How far did the Pacific plate move between the formation of the Kinmei Seamount and Kilauea? Assuming the plate has moved fairly steadily in that time, at what rate has it moved? (See "Rate of Spreading" on page 37.)

▶ Turn on and activate the **Absolute Motion (mm/yr)** theme.

This theme shows the speed and direction of plate motion relative to fixed locations (hotspots) for various points around the globe.

▶ Using the Identify tool [🛈], click on the brown arrow symbol just south of the Hawaiian islands. *(Important - see tip at left!)*
▶ Read the speed of the plate at that location in the Identify Results window.

19. How does the rate you calculated in the previous question compare to the absolute rate of motion in the middle of the Pacific Plate today?

Where to click?

When using the Identify tool on the Absolute (or Relative) Motion theme's arrows, clicking on the arrow symbol can be tricky. You need to click on the center of the arrow. Because the "arrowhead" is longer than the "shaft," the center is not where you expect it to be. The diagram below shows where to click.

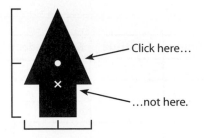

Click here…

…not here.

A (future) tale of two cities

Several of Earth's plate boundaries cut through heavily populated and economically important areas.

▶ Close the **Changing Plates** view and open the **Two Cities** view.

One of the most famous of these is the San Andreas Fault, a transform plate boundary that runs down the California coast. In time, the motion along this boundary will significantly change the face of the state and the fate of its two most famous cities.

▸ Use the Identify tool 🛈 to find the names of the plates on which San Francisco and Los Angeles are located.

20. On which plate is the city of San Francisco located? On which plate is the city of Los Angeles located?

▸ Turn on and activate the **Relative Motion (mm/yr)** theme.

▸ Use the Identify tool 🛈 to find the relative speed of the Pacific plate. This is how fast Los Angeles is approaching San Francisco.

21. How fast is the Pacific Plate moving northwest relative to the North American Plate?

▸ Use the Measure tool 📐 to measure the present distance (in meters) between San Francisco and Los Angeles. Read the distance in the status bar.

22. How far apart are Los Angeles and San Francisco today?

Can you find the time?

Since you know that...

   distance = speed × time

...you can rearrange the equation to find the time!

   time = distance / speed

23. At their current rate of relative motion, how many years will it take San Francisco and Los Angeles to be suburbs of one another? (Don't forget to change the distance to millimeters!)

24. There is lots of evidence that tropical palm trees used to grow in Antarctica and the southern parts of Chile and Argentina 30 million years ago. Explain how this could have happened when these places are so far from the tropics today.

Unit 3
Earthquake Hazards

In this unit, you will...

- *examine earthquake data to locate the largest and most destructive earthquakes,*

- *investigate deadly earthquake patterns throughout history,*

- *discover which factors contribute most to an earthquake's destructive potential, and*

- *explore the relationship between population, national wealth, and seismic risk.*

On December 7, 1988 a magnitude 6.9 earthquake struck northwestern Armenia, killing as many as 50,000 people and leaving over 500,000 homeless. Many historical buildings, such as this church in Leninakan, either collapsed or sustained severe damage. High, unsupported roofs make churches particularly vulnerable to earthquake damage.

Activity 3.1

The Great Lisbon Earthquake

Location Map

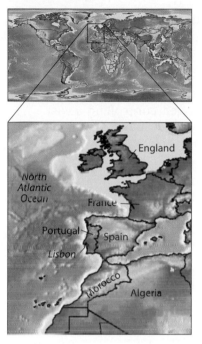

After reading the story of the earthquake that struck Lisbon, Portugal in 1755 (page 49), list and describe all of the earthquake-related hazards discussed. Feel free to add other hazards from your previous knowledge or experience with earthquakes.

Earthquake Hazards

1.

2.

3.

4.

5.

6.

7.

8.

Earthquake hazard examples (optional)

If you have access to a computer, you can see examples of these and other earthquake hazards. Be sure to add any new hazards you find to your list.

▶ Launch **ArcView GIS**.
▶ Choose **File ▶ Open…** and locate and open the **hazards.apr** file.
▶ Open the **Geological Hazards** view.
▶ To see examples of other earthquake hazards:
  • Turn on the **Hazard Links** theme and make it active.
  • Using the Hot Link tool, click on each of the orange earthquake hazard symbols (leaning buildings) on the map.
  • Read the caption for each picture, **then close its window**. There may be more than one picture for each link.

Hot tips for Hot Links

When you use the Hot Link tool,

• The theme containing the hot links must be active (the tool on the tool bar will be dark, not grayed out.)

• Click when *the* tip of the lightning bolt cursor is over the feature.

NO   YES!

Questions

1. Which of the hazards you listed caused the greatest amount of damage and cost the most lives in the Lisbon earthquake?

2. What do you think could have been done to reduce the loss of life and property in Lisbon? Explain.

Before 1755, Portugal was a major world power, controlling a vast global empire. In a weakened state after the earthquake, Portugal lost wars to Spain and England. Today, Portugal is the poorest country in Western Europe.

3. How might the world be different today if this earthquake had occurred somewhere else, or had been less severe?

4. If an earthquake of this magnitude occurred in a major US city today, how might it affect you?

Eyewitness account by Rev. Charles Davy

The Earthquake at Lisbon, 1755

"That was the year when Lisbon-town / Saw the earth open and gulp her down" From The One Hoss Shay, by Oliver Wendell Holmes.

THERE never was a finer morning seen than the 1st of November; the sun shone out in its full luster; the whole face of the sky was perfectly serene and clear; and not the least signal of warning of that approaching event, which has made this once flourishing, opulent, and populous city a scene of the utmost horror and desolation, except only such as served to alarm, but scarcely left a moment's time to fly from the general destruction.

It was on the morning of this fatal day, between the hours of nine and ten, that I was set down in my apartment, just finishing a letter, when the papers and table I was writing on began to tremble with a gentle motion, which rather surprised me, as I could not perceive a breath of wind stirring. Whilst I was reflecting with myself what this could be owing to, but without having the least apprehension of the real cause, the whole house began to shake from the very foundation, which at first I imputed to the rattling of several coaches in the main street, which usually passed that way, at this time, from Belem to the palace; but on hearkening more attentively, I was soon undeceived, as I found it was owing to a strange frightful kind of noise under ground, resembling the hollow distant rumbling of thunder. All this passed in less than a minute, and I must confess I now began to be alarmed, as it naturally occurred to me that this noise might possibly be the forerunner of an earthquake, as one I remembered, which had happened about six or seven years ago, in the island of Madeira, commenced in the same manner, though it did little or no damage.

Upon this I threw down my pen and started upon my feet, remaining a moment in suspense, whether I should stay in the apartment or run into the street, as the danger in both places seemed equal; and still flattering myself that this tremor might produce no other effects than such inconsiderable ones as had been felt at Madeira; but in a moment I was roused from my dream, being instantly stunned with a most horrid crash, as if every edifice in the city had tum-

bled down at once. The house I was in shook with such violence that the upper stories immediately fell; and though my apartment (which was the first floor) did not then share the same fate, yet everything was thrown out of its place in such a manner that it was with no small difficulty I kept my feet, and expected nothing less than to be soon crushed to death, as the walls continued rocking to and fro in the frightfulest manner, opening in several places; large stones falling down on every side from the cracks, and the ends of most of the rafters starting out from the roof. To add to this terrifying scene, the sky in a moment became so gloomy that I could now distinguish no particular object; it was an Egyptian darkness indeed, such as might be felt; owing, no doubt, to the prodigious clouds of dust and lime raised from so violent a concussion, and, as some reported, to sulphureous exhalations, but this I cannot affirm; however, it is certain I found myself almost choked for near ten minutes.

From *The Volcano's Deadly Work*, C. Morris, 1902

I hastened out of the house and through the narrow streets, where the buildings either were down or were continually falling, and climbed over the ruins of St. Paul's Church to get to the river's side, where I thought I might find safety. Here I found a prodigious concourse of people of both sexes, and of all ranks and conditions, among whom I observed some of the principal canons of the patriarchal church, in their purple robes and rochets, as these all go in the habit of bishops; several priests who had run from the altars in their sacerdotal vestments in the midst of their celebrating Mass; ladies half dressed, and some without shoes; all these, whom their mutual dangers had here assembled as to a place of safety, were on their knees at prayers, with the terrors of death in their countenances, every one striking his breast and crying out incessantly, Miserecordia meu Dios! . . . In the midst of our devotions, the second great shock came on, little less violent than the first, and completed the ruin of those buildings which had been already much shattered. The consternation now became so universal that the shrieks and cries of Miserecordia could be distinctly heard from the top of St. Catherine's Hill, at a considerable distance off, whither a vast number of people had likewise retreated; at the same time we could hear the fall of the parish church there, whereby many persons were killed on the spot, and others mortally wounded.

You may judge of the force of this shock, when I inform you it was so violent that I could scarce keep on my knees; but it was attended with some circumstances still more dreadful than the former. On a sudden I heard a general outcry, "The sea is coming in, we shall be all lost." Upon this, turning my eyes towards the river, which in that place is nearly four miles broad, I could perceive it heaving and swelling in the most unaccountable manner, as no wind was stirring. In an instant there appeared, at some small distance, a large body of water, rising as it were like a mountain. It came on foaming and roaring, and rushed towards the shore with such impetuosity, that we all immediately ran for our lives as fast as possible; many were actually swept away, and the rest above their waist in water at a good distance from the banks. For my own part I had the narrowest escape, and should certainly have been lost, had I not grasped a large beam that lay on the ground, till the water returned to its channel, which it did almost at the same instant, with equal rapidity. As there now appeared at least as much danger from the sea as the land, and I scarce knew whither to retire for shelter, I took a sudden resolution of returning back, with my clothes all dripping, to the area of St. Paul's. Here I stood some time, and observed the ships tumbling and tossing about as in a violent storm; some had broken their cables, and were carried to the other side of the Tagus; others were whirled around with incredible swiftness; several large boats were turned keel upwards; and all this without any wind, which seemed the more astonishing. It was at the time of which I am now speaking, that the fine new quay [wharf or reinforced shore where ships were loaded and unloaded], built entirely of rough marble, at an immense expense, was entirely swallowed up, with all the people on it, who had fled thither for safety,

and had reason to think themselves out of danger in such a place: at the same time, a great number of boats and small vessels, anchored near it (all likewise full of people, who had retired thither for the same purpose), were all swallowed up, as in a whirlpool, and nevermore appeared.

This last dreadful incident I did not see with my own eyes, as it passed three or four stones' throws from the spot where I then was; but I had the account as here given from several masters of ships, who were anchored within two or three hundred yards of the quay, and saw the whole catastrophe. One of them in particular informed me that when the second shock came on, he could perceive the whole city waving backwards and forwards, like the sea when the wind first begins to rise; that the agitation of the earth was so great even under the river, that it threw up his large anchor from the mooring, which swam, as he termed it, on the surface of the water: that immediately upon this extraordinary concussion, the river rose at once near twenty feet, and in a moment subsided; at which instant he saw the quay, with the whole concourse of people upon it, sink down, and at the same time every one of the boats and vessels that were near it was drawn into the cavity, which he supposed instantly closed upon them, inasmuch as not the least sign of a wreck was ever seen afterwards. This account you may give full credit to, for as to the loss of the vessels, it is confirmed by everybody; and with regard to the quay, I went myself a few days after to convince myself of the truth, and could not find even the ruins of a place where I had taken so many agreeable walks, as this was the common rendezvous of the factory in the cool of the evening. I found it all deep water, and in some parts scarcely to be fathomed.

This is the only place I could learn which was swallowed up in or about Lisbon, though I saw many large cracks and fissures in different parts; and one odd phenomenon I must not omit, which was communicated to me by a friend who has a house and wine cellars on the other side of the river, viz., that the dwelling-house being first terribly shaken, which made all the family run out, there presently fell down a vast high rock near it; that upon this the river rose and subsided in the manner already mentioned, and immediately a great number of small fissures appeared in several contiguous pieces of ground, from whence there spouted out, like a jet stream, a large quantity of fine white sand to a prodigious height. It is not to be doubted the bowels of the earth must have been excessively agitated to cause these surprising effects; but whether the shocks were owing to any sudden explosion of various minerals mixing together, or to air pent up, and struggling for vent, or to a collection of subterranean waters forcing a passage, God only knows. As to the fiery eruptions then talked of, I believe they are without foundation, though it is certain I heard several complaining of strong sulphureous smells, a dizziness in their heads, a sickness in their stomachs, and difficulty of respiration, not that I felt any such symptoms myself.

I had not been long in the area of St. Paul's when I felt the third shock, somewhat less violent than the two former, after which the sea rushed in again, and retired with the same rapidity, and I remained up to my knees in water, though I had gotten upon a small eminence at some distance from the river, with the ruins of several intervening houses to break its force. At this time I took notice the waters retired so impetuously, that some vessels were left quite dry, which rode in seven fathom water; the river thus continued alternately rushing on and retiring several times together, in such sort that it was justly dreaded Lisbon would now meet the same fate which a few years before had befallen the city of Lima; and no doubt had this place lain open to the sea, and the force of the waves not been somewhat broken by the winding of the bay, the lower parts of it at least would have been totally destroyed.

The master of a vessel which arrived here just after the 1st of November, assured me that he really concluded he had struck upon a rock, till he threw out the lead, and could find no bottom, nor could he possibly guess at the cause, till the melancholy sight of this desolate city left him no room to doubt of it. The two first shocks, in fine, were so violent that several pilots were of opinion the situation of the bar at the mouth of the Tagus was changed. Certain it is that one vessel, attempting to pass through the usual channel, foundered, and another struck on the sands, and was at first given over for lost, but at length got through. There was another great shock after this, which pretty much affected the river, but I think not so violently as the preceding; though several persons assured me that as they were riding on horseback in the great road leading to Belem, one side of which lies open to the river, the waves rushed in with so much rapidity that they were obliged to gallop as fast as possible to the upper grounds, for fear of being carried away.

I was now in such a situation that I knew not which way to turn myself: if I remained there, I was in danger from the sea; if I retired farther from the shore, the houses threatened certain destruction; and at last,

I resolved to go to the Mint, which being a low and very strong building, had received no considerable damage, except in some of the apartments towards the river. The party of soldiers, which is every day set there on guard, had all deserted the place, and the only person that remained was the commanding officer, a nobleman's son, of about seventeen or eighteen years of age, whom I found standing at the gate. As there was still a continued tremor of the earth, and the place where we now stood (being within twenty or thirty feet of the opposite houses, which were all tottering) appeared too dangerous, the courtyard being likewise full of water, we both retired inward to a hillock of stones and rubbish: here I entered into conversation with him, and having expressed my admiration that one so young should have the courage to keep his post, when every one of his soldiers had deserted theirs, the answer he made was, though he were sure the earth would open and swallow him up, he scorned to think of flying from his post. In short, it was owing to the magnanimity of this young man that the Mint, which at this time had upwards of two millions of money in it, was not robbed; and indeed I do him no more than justice in saying that I never saw any one behave with equal serenity and composure on occasions much less dreadful than the present.

Perhaps you may think the present doleful subject here concluded; but alas! the horrors of the 1st of November are sufficient to fill a volume. As soon as it grew dark, another scene presented itself little less shocking than those already described: the whole city appeared in a blaze, which was so bright that I could easily see to read by it. It may be said without exaggeration, it was on fire at least in a hundred different places at once, and thus continued burning for six days together, without intermission, or the least attempt being made to stop its progress.

It went on consuming everything the earthquake had spared, and the people were so dejected and terrified that few or none had courage enough to venture down to save any part of their substance; every one had his eyes turned towards the flames, and stood looking on with silent grief, which was only interrupted by the cries and shrieks of women and children calling on the saints and angels for succor, whenever the earth began to tremble, which was so often this night, and indeed I may say ever since, that the tremors, more or less, did not cease for a quarter of an hour together. I could never learn that this terrible fire was owing to any subterranean eruption, as some reported, but to three causes, which all concurring at the same time, will naturally account for the prodigious havoc it made. The 1st of November being All Saints' Day, a high festival among the Portuguese, every altar in every church and chapel (some of which have more than twenty) was illuminated with a number of wax tapers and lamps as customary; these setting fire to the curtains and timber-work that fell with the shock, the conflagration soon spread to the neighboring houses, and being there joined with the fires in the kitchen chimneys, increased to such a degree that it might easily have destroyed the whole city though no other cause had concurred, especially as it met with no interruption.

But what would appear incredible to you, were the fact less public and notorious, is that a gang of hardened villains, who had been confined and got out of prison when the wall fell, at the first shock, were busily employed in setting fire to those buildings which stood some chance of escaping the general destruction. I cannot conceive what could have induced them to this hellish work, except to add to the horror and confusion that they might, by this means, have the better opportunity of plundering with security. But there was no necessity for taking this trouble, as they might certainly have done their business without it, since the whole city was so deserted before night that I believe not a soul remained in it except those execrable villains and others of the same stamp. It is possible some among them might have had other motives besides robbing, as one in particular being apprehended (they say he was a Moor, condemned to the galleys), confessed at the gallows, that he had set fire to the king's palace with his own hand; at the same time glorying in the action, and declaring with his last breath that he hoped to have burnt all the royal family. It is likewise

generally believed that Mr. Bristow's house, which was an exceedingly strong edifice, built on vast stone arches, and had stood the shocks without any great damage further than what I have mentioned, was consumed in the same manner. The fire, in short, by some means or other, may be said to have destroyed the whole city, at least everything that was grand or valuable in it.

With regard to the buildings, it was observed that the solidest in general fell the first. Every parish church, convent, nunnery, palace, and public edifice, with an infinite number of private houses, were either thrown down or so miserably shattered that it was rendered dangerous to pass by them.

The whole number of persons that perished, including those who were burnt or afterwards crushed to death whilst digging in the ruins, is supposed, on the lowest calculation, to amount to more than sixty thousand; and though the damage in other respects cannot be computed, yet you may form some idea of it when I assure you that this extensive and opulent city is now nothing but a vast heap of ruins; that the rich and the poor are at present upon a level; some thousands of families which but the day before had been easy in their circumstances, being now scattered about in the fields, wanting every conveniency of life, and finding none able to relieve them.

Source - Eva March Tappan, ed., *The World's Story: A History of the World in Story, Song and Art, 14 Vols.*, (Boston: Houghton Mifflin, 1914), Vol. V: Italy, France, Spain, and Portugal, pp. 618-628.

Activity 3.2

Deadly earthquakes

Understanding where and when a major earthquake may strike is critical to assessing risk to life and property. It is also useful in developing plans to reduce that risk.

In this activity, you will investigate questions like "Where do the most damaging earthquakes occur?" You will also compare the locations of large earthquakes with the locations of deadly earthquakes and investigate changes in the distribution of deadly earthquakes over time.

▶ Launch **ArcView GIS**.

▶ Choose **File ▶ Open...** and locate and open the **hazards.apr** file.

▶ Open the **Earthquake Hazards** view.

This map shows the locations of nearly 8,000 earthquakes of magnitude 5 and greater. They are a representative sample of over 300,000 quakes recorded around the world since 1973.

Large earthquakes

Most damage occurs from earthquakes with magnitudes of 6.5 and greater. You will use a query to display only these large quakes.

▶ To show only the earthquakes of magnitude 6.5 or higher:

• Activate the **Earthquakes** theme.

• Choose **Theme ▶ Properties** and click the Query Builder 🔲 button.

• Enter the query (**[Magnitude] >= 6.5**) as shown below and click **OK** twice to display only the selected earthquakes.

To activate a theme, click on its name in the Table of Contents. Active themes are indicated by a raised border.

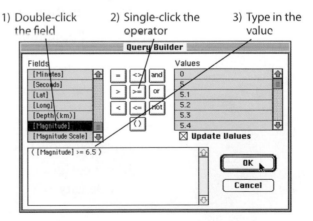

Now, only the large earthquakes are shown in the view.

To turn a theme on or off, click its checkbox in the Table of Contents.

▶ Turn on the **Plate Boundaries** theme.

▶ Use the Zoom In tool 🔍 and the Pan tool ✋ to examine the locations of the large earthquakes and the plate boundaries.

1. At which type of boundary are large magnitude earthquakes most common?

▶ Choose **View ▶ Full Extent** to display the entire map.

Deadly earthquakes

▶ Turn off the **Plate Boundaries** theme and turn on and activate the **Deadly Earthquakes** theme.

Deadly Earthquakes shows the locations of historical earthquakes in which 20 or more people died. Many of these events were magnitude 6 or greater, but quakes as small as magnitude 4 caused fatalities and significant damage. Prior to the early 1900s, there were no seismometers to record earthquakes, so magnitudes were estimated from written reports of earthquake damage.

▶ Turn the **Deadly Earthquakes** theme on and off and zoom in and out as needed to answer the following questions.

2. How do the locations of the deadly earthquakes compare to the locations of the large earthquakes?

▶ Choose **View ▶ Full Extent** to display the entire map.

Next, you will compare the locations of deadly earthquakes to the current distribution of the world's population.

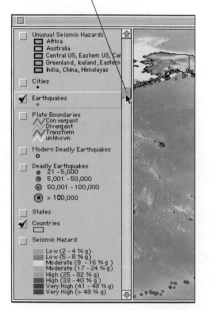
▶ Turn off the **Earthquakes** and **Topography** themes, and turn on the **Population Density** and **Deadly Earthquake** themes.

To create the **Population Density** theme, Earth's land area was first divided into a grid. Then, each square of the grid was filled with a color that represents the number of people living within the square.

▶ Scroll the Table of Contents down to see the **Population Density** legend.

3. Which color represents high population density? Which color represents low population density?

high density =

low density =

4. In general, how does the population density compare to the locations of deadly earthquakes?

5. Aside from magnitude and population density, what factors do you think might contribute to a higher damage and death toll in a large earthquake?

To turn a theme on or off, click its checkbox in the Table of Contents.

▸ Turn off the **Population Density** theme and turn on the **Topography** theme.

Deadly earthquakes throughout history

Next you will look at how the pattern of deadly earthquakes has changed over the past two centuries. You will perform queries to determine the number and locations of deadly earthquakes around the world during four time periods from 186 BC to the present.

▸ If necessary, choose **View ▸ Full Extent** to display the entire map.

▸ Activate the **Deadly Earthquakes** theme.

▸ Build a query to examine the deadly earthquakes prior to AD 500:

To activate a theme, click on its name in the Table of Contents.

• Choose **Theme ▸ Properties** and click the Query Builder button.

• Enter the query (**[Year] < 500**) as shown below and click **OK** twice to display only the selected earthquakes.

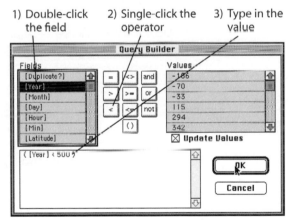

1) Double-click the field   2) Single-click the operator   3) Type in the value

Now only the deadly earthquakes that occurred before AD 500 are shown in the view.

▸ Click the Open Theme Table button to open the **Attributes of Deadly Earthquakes** table.

▶ To find death statistics for these earthquakes:
- Scroll the table to the right and select the **Deaths** field heading. (The field name is shaded when it is selected.)
- Choose **Field ▶ Statistics** to display statistics for the **Deaths** field and record them in the first column of the table below.
- Calculate the number of Deaths per Year (see example at left) and record it in the table.
- Scroll the table to the Location field and use the information to describe the regions where the deadly quakes occurred.

6. In the table below, record statistics for each of the time periods shown in the table columns.

NOTE – After recording earthquake statistics following each query, close the theme table and return to the Earthquake Hazards view. Perform a *new* theme properties query to complete the next column.

Calculating the number of deaths per year

Deaths per Year =
Number of deaths ÷ number of years

Example

467,960 deaths ÷ 686 years

rate = 682 deaths/year

Query statements

Enter your query statements exactly as shown in the table. If ArcView tells you that you have a syntax error when you click OK, re-enter the query statement again. Pay close attention to parentheses and brackets.

Time Period	186 BC–AD 499	AD 500–999	AD 1000–1499	AD 1500–2001
Query to enter	([Year] < 500)	([Year] >= 500) and ([Year] < 1000)	([Year] >= 1000) and ([Year] < 1500)	([Year] >= 1500)
Number of Years	686	500	500	501
Number of Quakes *(Statistics - Count)*				
Number of Deaths *(Statistics - Sum)*				
Deaths per Quake *(Statistics - Mean)*				
Deaths per Year *(Calculate - see example)*				
Region(s) *(from the Location field in the Deadly Earthquakes attribute table)*				

7. Describe how the locations of the deadly earthquakes have changed over time.

8. From the table, how has the number of deadly earthquakes changed over time?

9. How might cultural differences have influenced the pattern of deadly earthquakes shown here? Give an example.

What do you mean by "cultural differences"?

The deadly earthquake data were compiled recently by scientists and historians. To be listed, a deadly earthquake not only had to have occurred, but it must have been recorded by someone. Think about how different cultures pass along their history to future generations.

10. On the graph below, plot the average number of **deaths per year** for each time period.

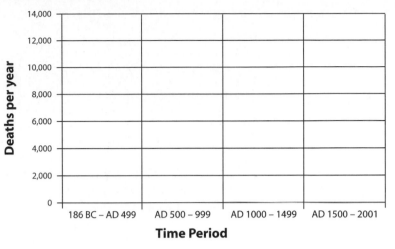

11. On the graph below, plot the average number of **deaths per earthquake** for each time period.

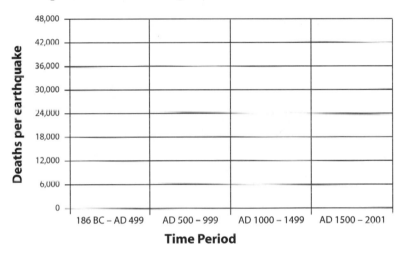

12. How have the number of deaths per year and average deaths per earthquake changed over time?

13. How might changes in the population distribution and improvements in earthquake detection contribute to these patterns?

14. Do these data show that the number of large earthquakes has been increasing over the past 2,000 years? Explain.

Activity 3.3

Seismic hazards

Hazards versus risks

The words *hazard* and *risk* are often used interchangeably, but they mean distinctly different things. A *hazard* is a potentially dangerous event or process. *Risk* is the potential loss of life, property, or production capacity due to a hazard. Geological hazards exist even if lives and property are not endangered. Risk, on the other hand, depends on how those hazards could affect human activities. In general, geologic hazards can't be controlled by humans. However, we can do much to minimize risks through wise land use, timely warnings, and community preparedness.

The idea of **acceptable risk** means being willing to live with a certain amount of risk in exchange for some economic or social benefit. Government agencies often decide how much risk is acceptable, balancing risks with economic, social, and political factors. For example, zoning laws may limit population density or restrict hospital construction in hazardous areas.

Most geological events occur in sparsely populated areas, and death tolls are low. Even small events can have deadly results, however, if they strike populated areas without warning. The worst situation is that of a large, unexpected event in a densely populated area.

History is the key

Depending on its size, type, and duration, a geological event can affect areas up to hundreds of kilometers from the its source. In the case of tsunamis, damaging effects can be global in extent. We can identify areas most likely to be affected by geologic hazards through a detailed study of an area's geologic history. Knowing the type and extent of previous activity, the geological setting, population density, and local construction practices, we can identify hazardous zones.

Recurrence intervals

If plates move steadily, earthquakes and volcanic eruptions should occur fairly regularly, given that the plate motions cause these events.

A recurrence interval is the time expected between two similar geologic events generated by the same geological feature. For example, if a magnitude 6 earthquake occurs on the same segment of a fault five times in 100 years, it has a recurrence interval of 20 years.

Although the historical record of earthquakes and volcanic eruptions is incomplete, it allows us to estimate how frequently major events occur in an area. By looking at the past, geologists can make educated guesses about what to expect in the future.

Unfortunately, people tend to view risk in terms of their lifetimes. This can be dangerous, because recurrence intervals for major earthquakes and volcanic eruptions can range anywhere from 20 years to as long as 20,000 years!

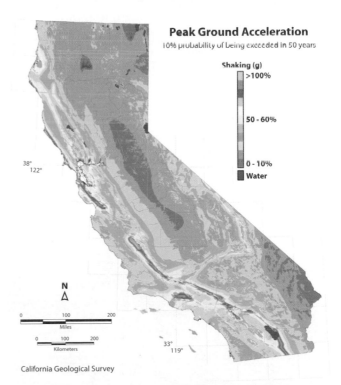

Peak Ground Acceleration
10% probability of being exceeded in 50 years

Shaking (g)
>100%

50 - 60%

0 - 10%

Water

California Geological Survey

This seismic risk map of California shows areas most likely to experience ground shaking in the next 50 years. Geological risk maps often include information about the intensity, extent, and expected frequency of a particular type of hazard event.

Seismic risk factors

Every day, hundreds of earthquakes occur somewhere on Earth. These earthquakes form identifiable patterns that outline tectonic plate boundaries. A significant number of these events occur on or near heavily populated coastlines. Some large earthquakes kill or injure thousands, while others go virtually unnoticed. So, what makes an earthquake deadly?

Three main factors influence how devastating an earthquake will be: (1) the intensity of ground shaking; (2) the population density of the region; and (3) economics or the wealth of the region.

Intensity of ground shaking

Ground shaking is controlled by the magnitude, depth, and distance from the earthquake, the direction of slip on the fault, and local conditions such as soil type and topography.

Magnitude—When rock under stress suddenly breaks and moves during an earthquake, it releases a tremendous amount of energy. Magnitude describes the size or strength of an earthquake. The magnitude increases when more rock is broken and slips along a fault. The amount of energy that is converted into seismic waves also increases with magnitude. These waves cause the ground shaking and damage that occur during earthquakes.

Depth and distance—The magnitude is not the whole story. As waves move away from the earthquake, their size and energy decrease. If an earthquake occurs at great distance or depth (~500 km away or below the surface) the waves generally cause little damage.

Earthquake magnitude scale

Source - http://www.seismo.unr.edu/ftp/pub/louie/class/100/magnitude.html

Magnitude	TNT For Seismic Energy Yield	Example	Earthquake Effects
–1.5	6 ounces	Breaking a rock on a lab table	Generally not felt but recorded by seismographs
1	30 pounds	Large blast at a construction site	
2	1 ton	Large quarry or mine blast	
3	29 tons	Often felt but damage rare	Often felt but damage rare
4	1,000 tons	Small nuclear weapon	
5	32,000 tons	Minor/moderate damage	Minor/moderate damage
6	1 million tons	Double Spring Flat, NV quake, 1994	Can be destructive in areas up to 100 kilometers across
6.5	5 million tons	Northridge, CA quake, 1994	
7	32 million tons	Hyogo-Ken Nanbu, Japan quake, 1995	Major earthquake. Can cause serious damage over larger areas
7.5	160 million tons	Landers, CA quake, 1992	
8	1 billion tons	San Francisco, CA quake, 1906	Great earthquake. Can cause major damage in areas hundred of kilometers across
8.5	5 billion tons	Anchorage, AK quake, 1964	
9	32 billion tons	Chilean quake, 1960	
10	1 trillion tons	San Andreas-type fault circling Earth	Theoretical quake (equal to Earth's daily "dose" of solar energy)
12	160 trillion tons	Fault through center of Earth	

In 1935 Dr. Charles Richter created a scale for measuring earthquakes in southern California. By measuring the amplitudes (heights) of seismic waves recorded on his seismograph, Richter was able to estimate the earthquake's energy. While Richter's scale applies only to earthquakes recorded in southern California, his work led to the development of a magnitude scale (moment magnitude) that can be used worldwide.

Near the earthquake, movement along a fault may rupture the ground surface. In addition to cracking building foundations and roads, surface ruptures can break gas, water, sewer, and communication lines. This creates additional health and safety hazards and hampers rescue efforts.

Soil type – Another critical factor in earthquake risk is the type of soil on which structures are built. Shaking in hard rock dissipates quickly, while loosely compacted sediments amplify seismic waves and intensify the ground shaking.

This process was dramatically illustrated during the 1985 Mexico City earthquake. The earthquake actually occurred 320 km west of the city, near the Pacific coast. There, the bedrock is very hard, so only minor damage and few deaths occurred. However, in Mexico City, which is built atop the soft sediments of an old lakebed, around 10,000

Karl V. Steinbrugge, Courtesy National Information Service for Earthquake Engineering, University of California, Berkeley

One of many buildings that collapsed in the magnitude 8.1 earthquake on September 19, 1985.

people were killed. The shaking triggered by the magnitude 8.1 quake was amplified by the loose soils, causing buildings to collapse or settle.

Soil conditions also played a major role in the 1989 Loma Prieta earthquake that affected San Francisco. An area of the city known as the Marina District sits on the edge of San Francisco Bay. The combination of loose fill and the high water table beneath this area enhanced the seismic shaking. The Marina District was one of the hardest hit areas of the city in the 1989 quake.

Human factors

Population density—An earthquake, no matter how powerful, doesn't pose a risk unless it occurs near where people live. The map (above right) shows the population density of California. Darker areas have a high density, while lighter areas have a lower density. Much of the population is clustered along the Pacific Coast, where San Diego, Los Angeles and San Francisco are located. Unfortunately, the coast is also the area where seismic activity occurs most frequently. (Compare it with the seismic risk map on page 51).

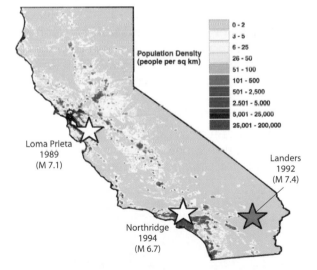

Population Density (people per sq km)	
	0 - 2
	3 - 5
	6 - 25
	26 - 50
	51 - 100
	101 - 500
	501 - 2,500
	2,501 - 5,000
	5,001 - 25,000
	26,001 - 200,000

Loma Prieta 1989 (M 7.1)

Landers 1992 (M 7.4)

Northridge 1994 (M 6.7)

When strong earthquakes occur where the population density is high, the effects can be catastrophic. Three recent California quakes illustrate the effect of population density on the overall impact of an earthquake. The two white stars represent the Loma Prieta and Northridge earthquakes. Together, these quakes killed almost 120 people and caused over 20 billion dollars in damage. The gray star shows the location of the Landers earthquake, the most powerful quake to hit California since 1952. Because of the low popu-

In San Francisco's Marina District, nine people died in building collapses and fires. This automobile was crushed under the third story of an apartment building. The first two floors of the structure are no longer visible.

lation density, the quake did only 100 million dollars in damage and killed just one person.

Human activities - Another critical factor is where people are and what they are doing when an earthquake strikes. Earthquakes that occur in the middle of the night offer some protection to people in wood frame homes, but may prove deadly to those in tall, poorly-constructed buildings made of unreinforced concrete.

The 1989 Loma Prieta earthquake struck just as a World Series baseball game between the two Bay Area teams was getting under way. Fortunately, many people who might otherwise have been driving home from work were watching the game. That lucky break probably saved hundreds of lives, because a large portion of a freeway normally packed with rush hour traffic collapsed during the earthquake. The upper deck of the double decker highway fell onto the lower deck, killing 42 people.

This section of Interstate 880 in Oakland, California collapsed in the 1989 Loma Prieta earthquake. Support columns in the two-tiered freeway failed, causing sections of the upper deck to "pancake" onto the lower deck.

The timing of an earthquake can also be disastrous. The January 26, 2001 Bhuj earthquake in India happened on Republic Day, a national holiday. When the quake struck, many citizens were in the streets, participating in or watching parades. Buildings collapsed onto the spectators, killing hundreds who might have otherwise been safe.

Economic factors

Gross Domestic Product (GDP)—The Gross Domestic Product is the total value of goods and services a country produces in a year. Usually, a country's GDP is divided by its total population to obtain the GDP per person, or per capita. Per capita GDP provides a good estimate of a country's wealth relative to other countries.

GDP per capita
- 500 - 1340
- 1341 - 2200
- 2201 - 3200
- 3201 - 4800
- 4801 - 6900
- 6901 - 9800
- 9801 - 13900
- 13901 - 19000
- 19001 - 25100
- 25101 - 34200

The world's per capita distribution of Gross Domestic Product. Wealthier countries are shown in red, poorer in blue.

A country's wealth has a dramatic effect on its ability to plan for, endure, and recover from the effects of a natural disasters. Affluent countries can afford to enforce stringent building codes and use superior construction materials. This reduces the chances of building collapse, which killed tens of thousands in recent quakes in Turkey and India.

Wealthier countries also have more modern search and rescue and health care systems to locate and treat the injured following earthquakes. Insurance networks and disaster relief programs help citizens and their communities rebuild after earthquakes.

Building construction—Building construction is a major factor in determining an earthquake's destructive capability. In earthquake prone regions, buildings must be able to withstand the horizontal shaking of an earthquake to prevent their collapse. Buildings should have strong connections between the walls, ceilings and floors to keep them from breaking apart.

Some of the most earthquake-safe buildings are single-story, wood frame houses with strong connections between the roof, walls and foundation.

Lindie Brewer, USGS

This sturdy wood frame house was located just 20 yards from the Landers earthquake fault scarp. While heavily damaged, the house did not collapse and its occupants, though shaken, escaped unharmed.

Wood is flexible and can move with the sway of the building. It may be damaged, but it generally does not collapse unless the roof is extremely heavy.

Concrete and brick buildings must be reinforced with steel to keep them from cracking and breaking apart. Steel is flexible yet strong and will bend rather than break like the concrete. On the other hand, failure in steel buildings most often occurs at weak welds between parts of the building.

Building damage often depends on the frequency of the ground motion—that is, how fast the ground vibrates. Damage can be particularly severe if the frequency of ground motion matches the natural vibration frequency of the structure, causing the building to resonate.

Resonance is similar to pushing a child on a swing. If you push at the right time, the swing goes higher and faster. If you push at the wrong time, the swing

E.V. Leyendecker, National Bureau of Standards

This steel frame building in Mexico City collapsed due to resonant ground shaking. The taller concrete building behind it suffered little damage.

slows down and ultimately stops. When seismic waves shake a building at its natural frequency, the building resonates and can quickly self-destruct.

Tall buildings, bridges, and other large structures respond more to low-frequency shaking, while small structures respond more to high-frequency shaking. During the 1985 Mexico earthquake, most of the buildings that collapsed were between seven and fifteen stories tall. The low-frequency ground vibrations of the earthquake matched the natural vibration frequency of buildings in this height range. Buildings shorter than seven stories or taller than fifteen stories sustained far less damage.

Other hazards

Fire—Broken gas and power lines often start fires after major earthquakes. These fires can easily spread out of control, particularly if broken water mains limit or cut off the water supply. Fires contributed significantly to damage and deaths following the 1906 and 1989 San Francisco earthquakes and the 1995 earthquake in Kobe, Japan.

Landslides—In mountainous areas, earthquakes can trigger landslides. The sudden and total devastation caused by landslides can produce extremely high death tolls. On May 21, 1970, a magnitude 7.9 earthquake at Mt. Huarascán, Peru triggered a rock and snow avalanche that buried the towns of Yungay and Hanrahirca. In all, the landslide killed over 20,000 people and caused more than $250 million dollars in damage.

Tsunamis—Earthquakes occurring in or near oceans can create seismic sea waves, or *tsunamis*. A large tsunami can travel great distances across an

The 1970 Mt. Huarascán landslide consisted of nearly 100 million cubic meters (80 million cubic yards) of water, mud, and rocks. A total of 67,794 people were killed by combined effects of the earthquake and landslide.

ocean with a minimal loss of energy. Thus, tsunamis are capable of causing destruction far from their source. Tsunamis and their hazards are covered in more detail in the Tsunami Hazards activity.

Hazardous spills—The growing use and storage of hazardous materials increases the potential risk from earthquakes. Damage to storage tanks, pipelines, and transportation systems for these materials could have catastrophic results that last well beyond the immediate effects of the earthquake.

Panoramic view of San Francisco on April 18, 1906 shows the city in flames following the magnitude 8.4 earthquake. The death toll was more than 3,000 from all causes, and damage was estimated at $500,000,000 in 1906 dollars (about $10 billion in 2000 US dollars).

Questions

1. What is the difference between a geologic hazard and a geologic risk?

2. California and Alaska both have very strong earthquakes. Why might the earthquake risk be much higher in California than it is in Alaska?

3. If a major earthquake occurs four times on a segment of a fault in 100 years, what is the recurrence interval of major earthquakes for that fault segment?

4. Why was so much damage done in Mexico City in 1985 by an earthquake that originated 320 km (200 miles) away?

5. Why was the death toll from the January 2001 earthquake in Bhuj, India unusually high?

6. What is "Per Capita GDP," and how might it affect the number of deaths in an earthquake?

7. Would you prefer to be in a modern steel frame skyscraper, a 3-story brick office building, or a 1-story wood frame house during an earthquake? Explain why.

8. Why, in some earthquakes, does most of the damage occur to buildings of a certain height, while taller and shorter buildings receive little damage?

Activity 3.4

Seismic risk and society

In this activity, you will explore the factors that influence seismic hazard and seismic risk around the world.

> ▶ If necessary, launch the ArcView GIS application, then locate and open the **hazards.apr** file.
> ▶ Open the **Earthquake Hazards** view. Turn off the **Earthquakes** and **Topography** themes.
> ▶ Turn on the **Seismic Hazard** theme.

The **Seismic Hazard** theme shows the maximum shaking, or ground acceleration, that an area could experience over the next 50 years. There is only a 10% probability that the maximum level of shaking will occur, so the map represents the worst case expected for the area. The hazard units are expressed as percentages of **g**, the acceleration of gravity (9.8 m/s²). The higher the ground acceleration, the more likely there will be structural damage and, as a result, injuries and deaths.

To turn a theme on or off, click its checkbox in the Table of Contents.

How likely is it?

When talking about hazards and risks, we usually talk about *probability*—how likely it is that some event will happen in a given period of time. In this case, the seismic hazard is defined as a 10% probability of some maximum ground acceleration over 50 years. According to the seismic hazard legend, an area classified as a Very High seismic hazard is rated at over 40% **g**. Thus, such an area has a 10% probability of experiencing an earthquake with shaking of 40% **g** (3.9 m/s²) or higher in the next 50 years.

To activate a theme, click on its name in the Table of Contents. Active themes are indicated by a raised border.

Seismic hazards, plate tectonics, and fault systems

As you've seen in your investigation of earthquakes and plate boundaries, large earthquakes are often associated with plate margins. Next, you will query the **Seismic Hazard** theme to see if this relationship holds true for earthquake hazards.

> ▶ Activate the **Seismic Hazard** theme.
> ▶ Choose **Theme ▶ Properties**.
> ▶ Click the Query Builder button 🔧 and enter a query to identify areas whose seismic hazard is Moderate to Very High. Enter the query expression (**[Hazard] = "Moderate") or ([Hazard] = "High") or ([Hazard] = "Very High")**. Click **OK** twice.

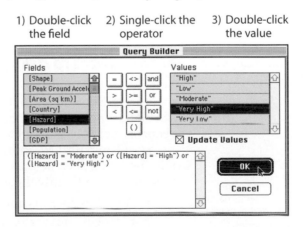

Now only the areas of moderate to very high seismic hazard are visible.

> ▶ Turn off the **Earthquakes** theme and turn on the **Plate Boundaries** theme.

▸ Use the zoom tools to examine seismic hazard areas near the plate boundaries. (You may need to zoom in close to examine these areas.)

1. Which plate boundary types are most commonly associated with areas of moderate to very high seismic hazard?

Seismic hazards far from plate boundaries

Not all seismic hazards are found near plate boundaries.

▸ Turn on and activate the **Unusual Seismic Hazards** theme.

This theme shows areas where seismic hazards exist far from recognized plate boundaries.

▸ Using the Identify tool 🛈, click in each of the five unusual seismic hazard regions. Read the information in the Identify Results window, then close the window.

2. Complete the table by filling in the name of the region that matches each seismic hazard source.

Idea for research

You may want to investigate some of these regions as a long-term research project.

Region	Source of seismic hazard
	Collision of two continents beginning 40 my ago. The broad zone of deformation is due to the fact that both continents are made of low density continental rock that cannot be subducted or destroyed.
	Tsunamis (seismic sea waves) originating at local plate boundaries and internal deformation of the plate related to ancient tectonics that formed the continent.
	Rifting - the continent is breaking apart, or rifting, along a newly-forming plate boundary.
	Glacial rebound - continents are rising in response to melting ice sheets 11,000 years ago. This causes small to moderate earthquakes.
	Seismic activity is related to weaknesses in the plates at ancient plate boundaries.

Glacial rebound

During the last ice age, around 20,000 years ago, thick ice sheets covered much of the northern hemisphere. The weight of the ice made the continents sink deeper into the mantle. As the ice sheets melted, the continents rose, or rebounded.

Though most of the ice melted long ago, glacial rebound continues today in places such as Greenland, Iceland, and eastern Canada, producing small amounts of seismic activity.

▸ Turn off the **Plate Boundaries** and **Unusual Seismic Hazards** themes.

Next, you will explore the difference between seismic hazard and risk.

From hazard to risk

Seismic hazard and seismic risk are related, but have very different meanings. Seismic hazard is the probability that ground shaking of a particular intensity will occur in a region, as determined from historic records. Seismic risk reflects society's ability to successfully endure an area's seismic hazard. In this section, you will investigate the relationship between seismic hazard and seismic risk. First, you will narrow your selection to only those areas whose seismic risk is "Very High."

▸ Activate the **Seismic Hazard** theme.

▸ Perform another theme properties query using the statement (**[Hazard] = "Very High"**).

▸ Click the Open Theme Table button 📧 to open the **Seismic Hazard** theme's attribute table.

▸ Scroll across the table to the **Country** field.

To activate a theme, click on its name in the Table of Contents. Active themes are indicated by a raised border.

Summary operations

Field	Summarize by
Population	Sum
Area (sq km)	Sum
GDP	First

Creating a summary table

When you create a summary table, you will have to specify a location for ArcView to store the data. Click the **Save As...** button and navigate to the filesystem location where you will save the data. Make sure you can write to the location you choose. This is especially important if you are running this activity from a CD-ROM or a lab fileserver; ArcView may default to the location of the project file itself, and you will likely be unable to write files to that location.

If you cannot create the summary table

From the Project window, select **Project ▶ Add Table....** With the Add Table dialog box, navigate to the Hazards project folder and open the Seismic Hazard Summary Table, summary.dbf.

What is GDP?

Gross Domestic Product (GDP) is the total amount of money that changes hands in a country in a given year. Per Capita GDP is simply this total amount of money divided by the number of number of people living in the country.

There are two different ground acceleration ranges labeled "Very High," so some countries have two records (rows) in the table. To compare countries, you need to combine these into a single record for each country. You can do this by creating a summary table.

▶ Select the **Country** field and click the Summarize button Σ.

▶ In the Summary Table Definition dialog box, choose *Population* in the **Field** pop-up menu and *Sum* in the **Summarize by** operation pop-up menu, then click **Add**. Repeat for the other fields and operations listed in the table at left.

▶ Click **OK** when you have entered all three summary operations as shown below. A new table called a *summary table* will open.

See "**Creating a summary table**," at left.

3. Complete the table below using data from the summary table.

Country [Country]	Data for very high seismic risk zones			Per Capita GDP 1998 US dollars [First_GDP]
	Population [Sum_Population]	Area [Sum_Area (sq km)]	Population density Population / Area	
China				
Japan				
Taiwan				
Turkey				
United States				

Ranking countries within the highest risk areas

Each factor in the above table contributes to the overall seismic risk of a region. A large population or high population density increases the risk of death or injury from an earthquake. A country's wealth, as indicated by its per-capita Gross Domestic Product (GDP), is also important in determining seismic risk. Affluent countries are better able to plan for and respond effectively to earthquakes.

4. Using the data in the table above, rank the countries for each risk factor in the table below. Calculate the Total Risk Ranking for each country by adding the risk values across the row for each country.

Country Name	Total population ranking (highest = 5 lowest = 1)	Population density ranking (highest = 5 lowest = 1)	GDP per capita ranking (lowest = 5 highest = 1)	Total Risk Ranking
China				
Japan				
Taiwan				
Turkey				
United States				

Next you will compare this theoretical risk ranking to the effects of recent earthquakes in each of these countries.

The following table illustrates the deaths and damage caused by five recent major quakes.

Country	Location	Year	Magn-itude	Deaths	Buildings Destroyed	Damage (billion $US)
China	Tangshan	1976	7.6	>240,000	180,000	5.6
Japan	Kobe	1995	6.8	5,502	100,000	150
Taiwan	Chichi	1999	7.6	2,400	10,000	14
Turkey	Izmit	1999	7.4	19,000	35,000	6.5
United States	Northridge	1994	6.7	57	7,000	44

5. How well does the risk factor you calculated in the risk matrix table compare to the actual earthquake damage and casualty figures for each of these countries? Discuss both the number of casualties and the total dollar damage.

Earthquake deaths and wealth

A country's per capita GDP provides a good measure of its wealth. In the risk matrix table, we assumed that affluent countries are better prepared for earthquakes and better equipped to deal with their effects.

Now you will use ArcView to test this assumption by looking at deadly earthquakes since 1950. You will examine how death and damage vary depending on whether the earthquake occurred in a wealthy country (GDP >= $10,000) or a poor country (GDP < $10,000).

To turn a theme on or off, click its checkbox in the Table of Contents.

To activate a theme, click on its name in the Table of Contents. Active themes are indicated by a raised border.

▶ Turn on the **Modern Deadly Earthquakes** theme.

This theme is a subset of the **Deadly Earthquakes** theme and shows the deadly earthquakes that have occurred since 1950.

▶ Activate the **Seismic Hazard** theme.

▶ To display all of the seismic risk zones again choose **Theme ▶ Properties**, click the **Clear** button, then click **OK**.

▶ Click the Query Builder button 🔍 on the button bar *(NOT the query builder button in the Theme Properties dialog box!)*, enter the query statement [GDP] >= 10000, click **New Set**, then close the Query Builder window.

This query selects countries whose per-capita GDP is at least $10,000 per year. These countries are considered wealthy, because they are well above the global mean per capita GDP of $7,500 dollars per year.

▶ Activate the **Modern Deadly Earthquakes** theme.
▶ Use the Select By Theme operation to select the deadly earthquakes that are within 150 km of any of the wealthy seismic hazard zones.
  • Choose **Theme ▶ Select By Theme**.
  • Configure the Select By Theme dialog box to read "Select features of active themes that **Are Within Distance Of** the selected features of **Seismic Hazard**" and enter **150** for the selection distance. Click the **New Set** button.

Select By Theme
Select features of active themes that

Are Within Distance Of ▼	New Set

the selected features of

Seismic Hazard ▼	Add to Set
	Select from Set

Selection distance:

150 km	Cancel

To open a theme table:
• activate the theme
• click the Open Theme Table button ⊞

▶ Open the **Modern Deadly Earthquakes** theme table. Scroll across and select the **Deaths** field and choose **Field ▶ Statistics**.
▶ Read the average (**Mean**) number of deaths per quake.

6. What is the average number of deaths per quake in countries with a high per capita GDP (over $10,000)?

▶ Close the statistics and theme table windows.
▶ Repeat this process to find the average number of deaths per quake in poor countries.

To activate a theme, click on its name in the Table of Contents.

  • Activate the **Seismic Hazard** theme and open its theme table.
  • Click the Switch Selection button ⊡. This will select all the countries whose per-capita GDP is below $10,000 per year.
  • Close the theme table windows.
  • Activate the **Modern Deadly Earthquakes** theme.
  • Perform the Select By Theme operation again, using the same settings you used above.
  • Open the **Modern Deadly Earthquakes** theme table and calculate statistics for the **Deaths** field.

7. What is the average number of deaths per quake in countries with a low per capita GDP (under $10,000)?

8. Describe how a country's GDP might affect the number of injured or killed.

9. Summarize the factors that increase seismic risk in a country.

Unit 4
Volcano Hazards

In this unit, you will...

- *examine records of historical volcanoes,*
- *use the Volcanic Explosivity Index (VEI) to categorize volcanoes,*
- *investigate the effect of major eruptions on climate, and*
- *explore the effects of the most explosive volcanoes in history.*

Lyn Topinka, USGS

Mount St. Helens viewed from the north, showing Spirit Lake and the new summit crater. The lateral blast from the May 18, 1980 eruption damaged or destroyed over four billion board feet of usable timber, enough to build 150,000 homes.

Activity 4.1

The tragedy of Mont Pelée

Location Map

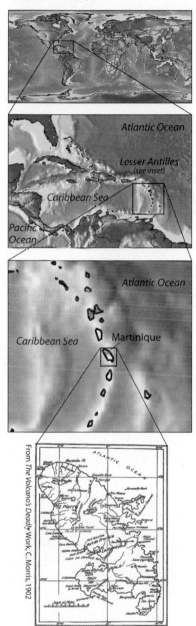

From *The Volcano's Deadly Work,* C. Morris, 1902

To use the Hot Link tool, position the *tip* of the lightning bolt cursor over the feature and click.

Like this   Not like this

After reading the account of the volcanic eruption that destroyed the town of St. Pierre on the island of Martinique in 1902 (page 73), list and describe all of the volcano-related hazards discussed. Feel free to add other hazards from your previous knowledge or experience with volcanoes.

Volcano hazards

1.

2.

3.

4.

5.

6.

7.

8.

Volcano hazard examples (optional)

If you have access to a computer, you can see examples of these and other volcanic hazards. Be sure to add any new hazards you find to your list.

▸ Launch **ArcView GIS**.
▸ Choose **File ▸ Open…** and locate and open the **hazards.apr** file.
▸ Open the **Geological Hazards** view.
▸ To see examples of other earthquake hazards:
  • Turn on and activate the **Hazard Links** theme.
  • Using the Hot Link tool [⚡], click on each of the brown volcano hazard symbols ▲ (brown volcanoes) on the map.
  • Read the caption for each picture, **then close its window**. There may be more than one picture for each link.

Questions

1. In St. Pierre, which of the hazards you listed caused the greatest amount of damage and cost the most lives?

2. Do you think anything could have been done to reduce the loss of life and property in St. Pierre? Explain.

3. What warning signs did the people of St. Pierre have of the coming eruption? Why do you think so many people ignored the signs?

4. Why do you think people build farms and cities so close to volcanoes?

5. Would you rebuild the city of St. Pierre? Explain why or why not.

Eyewitness accounts
The Destruction of St. Pierre, 1902

Volcanoes in the Caribbean

The islands of the Lesser Antilles are the surface expression of a volcanic arc above a subduction zone. For the past 40 million years, the Atlantic plate has been slowly plunging beneath the Caribbean plate, building this chain of tropical islands.

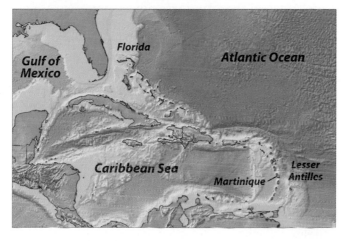

The rock and sediment of the ocean floor contain significant amounts of water. As the Atlantic plate carries these rocks downward into the mantle, the increasing heat and pressure releases the trapped water. At a depth of about 100 km (60 miles), this combination of heat, pressure, and water vapor cause the subducting plate to melt. The molten rock—magma—is much lighter than the surrounding mantle, so it rises by buoyancy forces toward the surface. Near the surface, the magma accumulates in reservoirs known as magma chambers. Occasionally, the pressure in the chamber increases, forcing magma to the surface in a volcanic eruption.

The shape and explosive nature of these volcanoes is a direct result of the magma's chemical content. At subduction zones, minerals with relatively low melting points such as silica (quartz) melt, while minerals with higher melting points remain solid. Magmas high in silica are very viscous; that is, they do not flow easily. Rather, the water vapor and other gases in the magma tend to build pressure until it is released explosively. This type of eruption forms steep-sided cones called stratovolcanoes that can reach thousands of meters in elevation. Mt. Rainier (outside Seattle), Mt. Fuji (in Japan), Popocatepetl (towering

over Mexico City) and Mt. Etna (on the island of Sicily) are all stratovolcanoes, and all are located at subduction zones.

The tropical islands of the West Indies, from St. Kitts and Montserrat in the north to St. Vincent and Grenada in the south, are stratovolcanoes that occasionally erupt with spectacular violence. The first recorded eruption in the West Indies occurred in about 1660 at La Soufrière on Montserrat. Since that time, dozens of eruptions have been observed and activity continues with regularity today. The deadliest eruption of the 20th century occurred in the West Indies on the island of Martinique.

Mt. Pelée and the island of Martinique

The island of Martinique sits at the midpoint of the Lesser Antilles island arc. For several hundred years the island had been a stopping point for pirates and buccaneers, but by the start of the 20th century it was home to over 160,000 residents. The largest city, the port of St. Pierre, was a thriving community of over 25,000. Spread out along a gently curving beach at the foot of cone-shaped Mont Pelée, it billed itself as the "Paris of the West Indies." Famous for its rum, St. Pierre had a distillery filled with barrels of the potent spirit.

From *The Volcano's Deadly Work*, C. Morris, 1902

Mont Pelée, rising several miles to the north of St. Pierre, was draped in a thick cover of jungle. At its summit, a bowl-shaped crater held a lake, *Lac des Palmistes*. Below the summit was a second crater, breached by a V-shaped canyon aimed like a rifle sight at St. Pierre, 6 km (3.8 mi) away. Though it was well known that Mont Pelée was a volcano, having produced minor eruptions in 1792 and 1851, the 1300-meter (4265-ft) peak was assumed by most to be dormant and therefore harmless. In fact, it was considered by many to be the island's benevolent "protector."

Mt. Pelée awakens

In late January 1902, Mont Pelée began to show signs that it was awakening from its long slumber. Smoke and steam began to puff from the summit. For the people of St. Pierre, it was more a source of wonder than alarm. This activity, called fumarole activity, gradually increased through the spring. Citizens occasionally noted the smell of sulfur, and on many days the mountains summit was covered with an ashen fog.

A. Lacroix

A later nuée ardente roars down the side of Mont Pelée in June 1902, passing over the already destroyed city of St. Pierre.

On April 23rd the townsfolk heard a loud explosion, and the summit of Mont Pelée billowed gray ash skyward in a minor eruption. Over the next few days, there were numerous explosions and tremors, and several times St. Pierre was dusted with fine volcanic ash. Conditions for the citizens of the town deteriorated; several times clouds of sulfurous gases, smelling of rotten eggs, permeated the city. The increasing activity of Mont Pelée caused the wildlife of the mountain to seek safer surroundings, while deadly snakes and swarms of insects invaded St Pierre and surrounding villages in search of food. Reports from diaries and letters describe how livestock screamed as red ants and foot-long centipedes bit them. Many livestock and an estimated 50 people, mostly children, died from snakebites.

On May 5th the water in the crater lake was heated to a boil and the crater rim failed. The hot water rushed down the canyons towards St. Pierre mixing with the recently fallen ash to create a volcanic mudflow called a lahar. The lahar destroyed everything in its path, including the premiere distillery where 23 workers were swept away and killed. When the lahar reached the ocean it created a local tsunami that flooded low-lying areas around the waterfront.

Obviously, life in the shadow of Mont Pelée was becoming unbearable. People were trying to leave the island, but passage aboard ships was difficult to secure. The cathedral was crowded with people waiting to make confessions. Seeking to comfort the populace, the governor sent a team to the summit to assess the danger. Only one scientist, the local high school science teacher, was in the group. The team delivered a positive report to the governor stating, "There is nothing in the activity of Mt. Pelée that warrants a departure from St. Pierre," and concluded that "the safety of St. Pierre is absolutely assured."

Unfortunately, the report was not accurate. On May 7th Pelée continued to be rocked with explosions, and people noted that two fiery "eyes" appeared near the summit. Above the summit hung a gray ash cloud filled with lightning. Marino Leboffe, captain of the Italian merchant ship *Orsolina*, knew what was coming. His home port was Naples, which lies in the shadow of Mount Vesuvius, a volcano that erupted with deadly force in 79 AD, completely burying the cities of Pompeii and Herculaneum. Leboffe ordered the *Orsolina* to leave port only half loaded with cargo and noted, "I know nothing of Mont Pelée, but if Vesuvius were looking the way your volcano looks this day, I'd get out of Naples; and I'm going to get out of here." Threatened with arrest for leaving port without clearance papers, Leboffe replied, "I'll take my chance of arrest, but I won't take any chances on that volcano."

The eruption on May 8th

Shortly before 8:00 am on May 8th a series of violent explosions rocked Mont Pelée. Thus began a huge volcanic eruption, first by sending a plume of gases and ash skyward, followed by a lateral eruption that sent a deadly superheated cloud of gases through the "V" notch directly towards St. Pierre. The eruption that went upward created a huge mushroom cloud that blocked out the early morning sun and was pierced by flashes of lightning. People 30 km from the volcano were immersed in darkness, and could not see even an arm's length away. This cloud reached an elevation of nearly 9 km (30,000 feet) in a matter of a few minutes.

Although spectacular, the vertical ash plume of ash was not the deadliest messenger from Pelée. The lateral eruption of superheated gases roared toward St. Pierre at over 160 km per hour (100 mph). The shock wave flattened brick buildings and ripped branches from trees. "Rubble walls three feet in thickness had been torn to pieces as if made of dominoes or kindergarten blocks." Not a single roof remained standing. It was reported

An artist's depiction of the May 8, 1902 eruption

that a three-ton cast iron statue was thrown 15 m (50 feet) from its pedestal. The nuée ardente, or "glowing cloud" of superheated gases and debris, glowed red, searing everything in its path and igniting fires throughout the city. At the distillery, rocks propelled by the cloud pierced centimeter-thick iron storage tanks with holes as large as 30 cm (1 foot) in diameter. Thousands of barrels of rum exploded, sending flaming rivers down the streets of the city and out to sea.

The eruptions were immediately followed by torrential rains. The runoff mixed with volcanic ash to form mudflows, called lahars, that swept down river valleys, filling them with debris and burying or sweeping houses off their foundations. The lahars flowed into the sea so suddenly that they generated large waves—tsunamis—that were observed around the Caribbean.

When the eruption finally subsided, more than 29,000 people had been killed. Only two residents of St. Pierre were reported to have survived, making this eruption the deadliest of the 20th century.

Stories of survivors

The force of the eruption was not limited to the city. In the harbor, the shock wave capsized steamships and raining pyroclastic debris set ships aflame. Chief Officer Ellery Scott of the Canadian steamship *Roraima* later told about his experience.

Chief Officer Ellery Scott

According to Scott, "The ship arrived at St. Pierre at 6 A.M. on the 8th. At about 8 o'clock, loud rumbling noises were heard from the mountain overlooking the town, the eruption taking place immediately, raining fire and ashes; lava running down the mountainside

with a terrific roar, sweeping trees and everything in its course. I went at once to the forecastle-head to heave anchor. Soon after reaching there, there came a terrible downpour of fire, like hot lead, falling over the ship and followed immediately by a terrific wave which struck the ship on the port side, keeling her to starboard, flooding ship, fore and aft, sweeping away both masts, funnel-backs and everything at once.

"I covered myself with a ventilator standing nearby, from which I was pulled out by some of the stevedores, and dragged to the steerage apartment forward, remaining there for some time, during which several dead bodies fell over and covered me. Shortly after, a downfall of red hot stones and mud, accompanied by total darkness, covered the ship. As soon as the downfall subsided, I tried to assist those lying about the deck injured, some fearfully burnt. Captain Muggah came to me, scorched beyond recognition. He had ordered the only life boat left to be lowered; but it was too badly damaged. From that time, I saw nothing of the captain; but was told by a man that the captain was seen by him to jump overboard. The man followed him in the water, and succeeded in getting the captain on a raft floating nearby, where he died shortly after.

The ruined city of St. Pierre, as it appeared on February 19, 1903.

"I gave all help possible to passengers and others lying about the deck in dying condition, most of whom complained of burning in the stomach. I picked up one little girl lying in the passageway dying, covered her over with a cloth, and took her to a bench nearby, where I believe she died. About 3 P.M. a French man-of-war's boat, the *Suchet*, came alongside and passed over the side about twenty persons, mostly injured, and myself and other survivors were taken to Fort de France. I afterwards saw the *Roddam* steaming out to sea, with her stern part on fire. The *Roraima* caught fire and was burning when I left her in the afternoon, the town and all shipping destroyed."

Overall, 48 of the 68 crew members and passengers died in the horrible ordeal, while on other ships the casualty rate was even higher.

Excerpted from *The Volcano's Deadly Work: from the fall of Pompeii to the destruction of St. Pierre*, by Charles Morris, 1902 (publisher unknown).

Ciparis

Although only two people survived the eruption in St. Pierre, several people lived in the surrounding communities and chronicled some remarkable tales. The most amazing story was that of Louis Auguste Sylbaris, also known as Ciparis. Ciparis was a robust 25-year-old who had a passion for drinking and brawling. In early April, he was arrested and jailed for wounding one of his friends with a sword during an argument. Jail time meant that Ciparis was required to labor in the service of the city, and he soon tired of the regimented life. Near the end of April, he escaped and partied all night with friends.

In the morning he turned himself in to the authorities, who sentenced him to one week of solitary confinement in the dungeon. The dungeon was next to a steep hillside, totally isolated from the outside world. On the morning of May 8, while Ciparis was waiting for his breakfast to be delivered, his cell became very dark. Gusts of hot air and steaming ash came through the cracks in his cell door. The heat became unbearable and he held his breath as long as he could. Finally, the heat began to subside and he slumped to the floor. He was horribly burned, but had some water in the cell to drink.

Ciparis managed to survive for four days before being rescued. An American journalist interviewed him shortly after his rescue and stated, "He had been more frightfully burned, I think, than any man I had ever seen." Ciparis eventually recovered, and was pardoned for his crimes on the basis of the miracle of his survival. Later, he joined the Barnum and Bailey Circus, and he was billed as the "Lone Survivor of St. Pierre."

Aftermath of the eruption

News of the deadly eruption was telegraphed to the world, and soon aid and supplies were rushed to Martinique. Along with this aid came scientists to study the volcano including Alfred Lacroix, often called the "father of modern volcanology. " It was Lacroix who coined the phrase "nuée ardente," which he described as a "lateral blast propelled down-slope by gravity."

On May 20th of 1902, Mont Pelée exploded again with a giant eruption, probably larger than that of May 8th. No one died in this event, mostly because there was no one left. Throughout the summer and fall there were many small eruptions, and on August 3rd another nuée ardente destroyed the village of Morne Rouge southeast of St. Pierre, killing 1,000 to 2,000 people.

In October of 1902 a lava dome began to rise out of the crater floor. This dome formed an imposing obelisk that has been described as the most impressive lava dome ever produced; it was 100-150 m (325-500 feet) thick at its base and soared to over 300 m (980 feet) above the crater floor. At times it rose at the remarkable rate of 15 meters per day! This obelisk, nicknamed "the tower of Pelée," glowed an incandescent red at night until it finally became unstable and collapsed into a pile of rubble in March of 1903.

A. Lacroix

View of the obelisk, or spire, rising 358 meters above the crater rim.

For a detailed account of the events surrounding the 1902 eruption, see *The Day the World Ended* by Gordon Thomas and Max Morgan Witts. (New York: Stein and Day, 1969.)

Activity 4.2

Deadly volcanoes

Volcanoes have the potential for affecting much larger areas and numbers of people than earthquakes. Mudflows generated by an eruption can travel hundreds of kilometers and ash can be carried by atmospheric currents all the way around the world.

In this activity, you will investigate different types of volcanic hazards and their effects on people and the landscape.

▸ Launch **ArcView GIS**.
▸ Choose **File ▸ Open...** and locate and open the **hazards.apr** file.
▸ Open the **Volcano Hazards** view.

Historical volcanoes

The **Volcanoes** theme shows the locations of all volcanoes that are known to have been active in the past 10,000 years. The **Volcanic Eruptions** theme contains information about known volcanic eruptions since 79 AD. These data are primarily gathered through written historical accounts as well as field investigations by geoscientists. The data are reliable for those regions where written history was preserved and sparse where oral history was the tradition.

▸ Use the zoom and pan tools to closely examine the relationship between volcanoes and plate boundaries.

1. With which type of plate boundary are the volcanoes most strongly associated?

To turn a theme on or off, click its checkbox in the Table of Contents.

To activate a theme, click on its name in the Table of Contents.

▸ Choose **View ▸ Full Extent** to see the entire view again.
▸ Turn off the **Volcanoes** and **Plate Boundaries** themes.
▸ Activate the **Volcanic Eruptions** theme.
▸ Click the Open Theme Table button 🗐 to open the **Volcanic Eruptions** theme table.
▸ Read the total number of volcanic eruptions in the tool bar.

read total here

2. How many volcanic eruptions have been recorded since 79 AD?

▸ Close the theme table window.

Volcanic Explosivity Index (VEI)

No single feature describes the size of a volcanic eruption. To compare the energy of eruptions, volcanologists developed a magnitude scale called the Volcanic Explosivity Index (VEI). The calculation factors in the height of the eruption plume, the distance ejecta travel, the duration of the blast, and the volume of material erupted. A VEI difference of 1 represents a difference in energy of approximately 10 times. For historical eruptions, VEIs range from 0-7, with 7 being the most explosive.

Which one's the biggest?

The only eruption in recent history with a VEI of 7 was the April 18, 1815 eruption of Mt. Tambora in Indonesia. To see a table of other historic eruptions and their intensities, click the Media Viewer button 📽 and choose **VEI Index**.

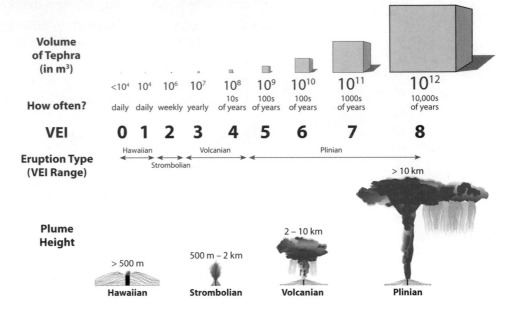

Now you will search for the largest historical eruptions.

▸ Build a query to select the volcanic eruptions with a VEI greater than four.
  • Click the query builder button 🔍.
  • Enter the query statement ([VEI] > 4), click the **New Set** button, and close the query builder window.
▸ Click the Open Theme Table button 📰 to open the **Volcanic Eruptions** theme table. Read the number of major eruptions from the tool bar.

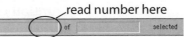

read number here

To calculate a percentage

Divide the number selected by the total number and multiply by 100. For example, if 145 of 1300 eruptions had a VEI greater than or equal to 4, the percentage of eruptions with a VEI >= 4 would be

$$\% = 145/1300 \times 100$$
$$= 0.112 \times 100$$
$$= 11.2$$

3. How many major eruptions (greater than VEI 4) have there been since 79 AD?

4. What percentage of the total eruptions were major ones (VEI > 4)?

5. If an average of 250 volcanic eruptions occur each decade, how many eruptions would have a VEI greater than 4?

> Choose **Edit ▶ Select None** to deselect the major eruptions, then close the theme table window.

Historical records: a window to the future?

Scientists routinely use records of the past to make educated guesses about events in the future.

6. What factors might affect the reliability of historical data and our estimation of future hazards? Explain.

Recurrence interval

Recurrence interval simply means how often an event happens, or the number of years between similar events. Calculate this by dividing the total number of years by the number of events that occurred in that amount of time.

Building a compound query

A compound query contains two or more statements of a *Field Name*, an *operator*, and a *Value* compared using a logical operator such as *and*.

To build this query,

- find and double-click **[Name]** on the Fields list
- click the = button
- type in the value **"St. Helens, Mt."**
- click the **and** button
- find and double-click **[VEI]** on the Fields list
- click the >= button
- type in the value **3**

The query should look exactly like this:

([Name] = "St. Helens, Mt.") and ([VEI] >= 3)

If you get an error message when you click **OK**, re-build your query from scratch and try again.

"Hey! This isn't how I did a query before!"

Before, you used a Theme Properties query, which identifies the features you are looking for by hiding the ones you *aren't* looking for. Now, you're doing a "regular" query, which shows the features you are looking for by highlighting them yellow, both in the map and in the theme table.

Historical data are often used to calculate recurrence intervals of major geologic hazards in a region. These are then used to estimate future levels of hazards. Recurrence can be calculated by dividing the number of years over which an event is repeated by the number of times the event occurred.

Next you will use a query to determine the recurrence interval of Mt. St. Helens eruptions with VEI ratings of 3 or greater.

> Build a compound query (see directions at left) to determine the number of times Mt. St. Helens has erupted with a VEI of 3 or greater.
>   - Activate the **Volcanic Eruptions** theme.
>   - Click the query builder button.
>   - Build the query **([Name] = "St. Helens, Mt.") and ([VEI] >= 3)**.
>   - Click the **New Set** button and close the query builder window.

> Click the Open Theme Table button to open the **Volcanic Eruptions** theme table. The number of eruptions found by your query is shown in the status bar.

read number here

> Click the Promote button to move the highlighted records to the top of the table. Scroll the table as needed to answer the following questions.

7. How many times has Mt. St. Helens erupted with a VEI of 3 or higher? Over what time interval did those eruptions occur? (Hint: The database lists eruptions only through the year 2001.)

8. What is the recurrence interval for eruptions of VEI >= 3 on Mt. St. Helens? How would you use this information in managing the future development and use of the area surrounding the volcano?

> Choose **Edit ▶ Select None** to deselect the Mt. St. Helens eruptions, then close the theme table window.

Range of volcanic hazards

On May 18, 1980 Mt. St. Helens erupted with a blast of VEI = 5 that deposited a blanket of ash over areas as far away as 400 km (250 mi). Lahars flowed down river valleys as far as 30 km (19 mi). Spokane, Washington, was significantly affected by the event. What would be the impact of a similar eruption at Mt. Rainier near Seattle, Washington?

Finding Seattle

Seattle, Washington is located near Puget Sound, on the US west coast.

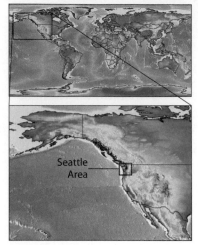

▶ To figure this out:

- Turn off the **Volcanic Eruptions** theme and turn on the **Volcanoes** theme.
- Turn on and activate the **Cities** theme.
- Use the Zoom In tool ⊕ to zoom in on the Seattle area. (See locator map, left.) The red circle shows the area within a radius of 100 km of Mt. Rainier.
- Using the Pointer tool ⊾, click on the shaded circle to select it. Four small black "handles" will appear around the circle when it is selected.
- Click the Select by Graphics button ⊞ to select the cities within 100 km of Mt. Rainier. They will be highlighted yellow.
- Click the Open Theme Table button ⊞ to open the **Cities** theme table.
- Scroll across the table to the **Area Population** field, click the field label to highlight it, then choose **Field ▶ Statistics**. The number of people affected is given as the **Sum**.

9. How many people living within 100 km of Mt. Rainier would be significantly affected if it erupted like Mt. St. Helens?

Finding Mt. Pinatubo

Mt. Pinatubo is located on the island of Luzon in the Philippines.

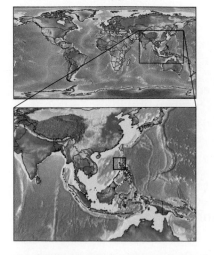

In 1991 Mt. Pinatubo in the Philippines erupted, causing widespread damage. The eruption was similar to the Mt. St. Helens eruption of 1980, only larger. With a VEI of 6, this larger blast affected an area around the volcano with a radius of about 500 km.

▶ Choose **Theme ▶ Clear Selected Features**.

▶ Choose **View ▶ Full Extent** to see the entire map.

▶ Use the locator map at the left to zoom in on Mt. Pinatubo in the Philippines. Repeat the steps above to determine how many people would be affected by another eruption of Mt. Pinatubo.

10. How many people would be significantly affected by another eruption of Mt. Pinatubo?

Activity 4.3

Volcanic hazards

Volcanoes are capable of affecting much larger areas and numbers of people than are earthquakes. Fortunately, they often provide warning signs of upcoming eruptions, allowing people to evacuate to safety. Still, there have been 35 volcanic eruptions in the past 500 years in which three hundred or more people were killed.

Volcanic hazards fall into two categories.

- Direct effects—eruption clouds, shock waves from the eruption blast, lava and pyroclastic flows, and volcanic gases.
- Indirect effects—lahars (mud flows of volcanic ash), flooding, tsunamis, and post eruption starvation.

Eruption clouds

Explosive eruptions blast rock fragments (tephra) and superheated gases into the air with tremendous force. The distance these fragments travel depends on their size—smaller particles go farther than larger fragments. Volcanic bombs, the largest fragments, generally land within 3 km of the vent. Particles less than 2 mm (0.08 in) across, called volcanic ash, can rise high into the air forming a huge, billowing eruption cloud or column.

Eruption columns grow rapidly and can reach heights of more than 20 kilometers (12 miles) in

Ash column of the July 22, 1980 eruption of Mount St. Helens.

less than 30 minutes. Large eruption clouds extend hundreds of miles downwind, resulting in ash fall over broad areas. Ash from the 1980 eruption of Mount St. Helens covered more than 50,000 square kilometers (20,000 square miles). Heavy ash fall can damage or collapse buildings. Even minor ash falls can damage crops, electronics, and machinery.

Volcanic ash clouds also pose a serious hazard to air travel. Each year, thousands of aircraft fly routes over volcanically active areas. During the past 15 years, about 80 commercial jets have been damaged by accidentally flying into ash clouds. Several have nearly crashed when ash coated the engines and caused them to fail. The drifting cloud can carry the hazard far from its source. In 1992, Chicago's O'Hare airport was forced to close for several hours due to a drifting ash cloud from Alaska's Mt. Spurr volcano, which is nearly 5,000 km (3,000 miles) from Chicago!

A heavy load of ash from the September 1994 eruption of the Rabaul Caldera in the Philippines damaged the roof of this house. Dry ash weighs 400 to 700 kg/m³ (880 to 1,545 lb/yd³). If ash becomes saturated with rainwater, its weight can double. In addition to the collapse risk, people trying to remove the ash face health and safety hazards.

Volcanic gases

At depth and under high pressure, magma can hold large amounts of dissolved gases. As the magma rises toward the surface and the pressure decreases, the gases bubble out and escape. This is similar to what happens when you open a can of soda.

Near the surface, the magma interacts with the surrounding rocks and ground or surface water to generate more gases. These gases may be released suddenly in explosions or seep slowly to the surface through cracks in the overlying rock.

The behavior of the magma is related to the amount of silica and water it contains. Magmas with a high silica content are generally more explosive.

Location	Source	% Silica	Explosivity
hot spots	lower mantle	50%	low
divergent boundaries	upper mantle	50%	low
convergent boundaries	upper mantle and crust	60–70%	medium to high

The table below shows volcanic gas compositions from three types of volcanic sources.

Gas	Kilauea hot spot 1170° C	Erta` Ale divergent plate 1130° C	Momotombo convergent plate 820° C
water vapor (H$_2$O)	37.1	77.2	97.1
carbon dioxide (CO$_2$)	48.9	11.3	1.44
sulfur dioxide (SO$_2$)	11.8	8.34	0.50
hydrogen (H$_2$)	0.49	1.39	0.70
carbon monoxide (CO)	1.51	0.44	0.01
hydrogen sulfide (H$_2$S)	0.04	0.68	0.23
hydrogen chloride (HCl)	0.08	0.42	2.89
hydrogen fluoride (HF)	-	-	0.26

Data from Symonds et al. 1994.

As you can see, much of the gas released by volcanoes is water vapor. Some water is dissolved in the magma itself, but most water vapor is produced when the magma comes in contact with ground water.

Carbon dioxide

Carbon dioxide is denser than air, causing it to accumulate in low areas. High concentrations of CO$_2$ (above 10%) are poisonous, but even lower concentrations are deadly because the CO$_2$ displaces oxygen required for respiration.

At California's Mammoth Mountain, the accumulation of CO$_2$ in the soil and in low areas is killing trees and small animals and poses a threat to humans unaware of the danger.

K. McGee, USGS

Trees killed by CO$_2$ near Mammoth Mountain, California.

Lake Nyos, in Cameroon, west-central Africa, sits in a large volcanic caldera. In 1986 a tremendous cloud of CO$_2$ suddenly escaped from the depths of the lake and flowed down a valley toward a nearby village. Before the dense gas dissipated, it had killed more than 1,700 people and livestock up to 25 km (16 miles) from the lake. This is not a unique event—similar cases have occurred elsewhere and will continue to occur.

J. Lockwood, USGS

Lake Nyos in western Cameroon, ten days after releasing a cloud of carbon dioxide gas that killed 1700 people in the valley below, at distances up to 27 km from the lake.

Ground water carries large amounts of dissolved CO$_2$ from the underlying magma into the bottom of the lake. Over time, the water becomes oversaturated with CO$_2$. If a disturbance causes some of this water to rise, the decreased pressure allows bubbles of CO$_2$ to form. The rising bubbles further disturb the water, triggering a chain reaction, resulting in a sudden, massive release of CO$_2$.

Air pollution—"vog" and "laze"

Steady eruptions, like those that occur at Hawaii's Kilauea volcano, can be significant sources of local air pollution. Under the right conditions, erupting sulfur dioxide (SO_2) gas combines with water vapor to form tiny sulfuric acid droplets. These interact with other volcanic gases, dust, and sunlight to produce volcanic smog, or "vog." Vog reduces visibility, causes health problems, damages crops, and corrodes metals.

When molten lava comes in contact with seawater, the intense heat produces clouds of lava haze, or "laze." More complex than simple clouds of steam, the heat breaks down the salt to form hydrochloric acid (HCl). The resulting cloud has a pH of 1.5–2.5 and can create significant health problems for downwind populations.

Pyroclastic flows descend the southeastern flank of Mayon Volcano, Philippines on September 23, 1984.

Lava from a 1993 eruption of Kilauea reacts with seawater to produce plumes of highly acidic "laze" (lava haze).

Pyroclastic flows

Pyroclastic flows—sometimes called nuées ardentes (French for "glowing clouds")—are hot, often glowing mixtures of volcanic fragments and gases that flow down the slopes of volcanoes. These flows can reach temperatures of 800°C (1,500°F) and move at speeds up to 725 km/hr (450 mph), flattening and burning everything in their path.

The eruption of Mount St. Helens on May 18, 1980, produced a sideways "lateral blast" that destroyed an area of 230 square miles. Trees were mowed down like blades of grass as far as 24 km (15 miles) from the volcano.

Lava flows and domes

Lava flows occur when molten rock (magma) flows or oozes out onto Earth's surface. Lava flows bury land and structures, start fires, and may block roads or streams.

How easily lava flows depends largely on its silica (silicon dioxide, SiO_2) content. Low-silica basaltic lava can form fast-moving (low viscosity) (10 to 30 miles per hour) streams or spread out in thin sheets many kilometers wide. Since 1983, basaltic lava flows from Hawaii's Kilauea Volcano have destroyed more than 200 buildings and cut off a nearby highway.

A lava flow from Kilauea blocks Hawaii's Highway 19 on its way to the sea.

In contrast, flows of higher-silica lava tend to be thick and sluggish (high viscosity), traveling only short distances from the vent. These thicker lavas often squeeze out of a vent to form irregular mounds called lava domes. Between 1980 and 1986, a lava dome at Mount St. Helens grew to a height of about 300 meters (1,000 feet). Lava domes seldom pose risks to humans.

Landslides

Volcanic landslides range from small surface movements of loose debris to collapses of the entire summit or sides of a volcano. These landslides are triggered when eruptions, heavy rainfall, or large earthquakes cause material to break free and move downhill.

Landslides are particularly common on steep composite or stratovolcanoes, which are often built up of layers of loose volcanic rock fragments. Heavy precipitation or groundwater can turn volcanic rocks into soft, slippery clay minerals, reducing the energy needed to trigger a slide.

The largest volcanic landslide in historical time occurred at the start of the May 18, 1980 Mount St. Helens eruption. In all, over 2.3 cubic kilometers (2/3 cubic mile) of the mountain were transported up to 24 kilometers (15 miles) downhill at speeds over 240 km/hr (150 mph). The slide left behind a hummocky (lumpy) deposit with an average thickness of 45 m (150 feet).

For years, geologists were puzzled by the strange, "lumpy" landscape north of Mt. Shasta in northern California shown in the picture below.

Ancient landslide from Mt. Shasta, California, shown in red, was 20 times the size of the 1980 Mt. St. Helens landslide.

After witnessing the eruption of Mt. St. Helens, with its tremendous landslide and the hummocky terrain it left behind, Mt. Shasta now made sense.

A massive landslide had occurred about 350,000 years ago when nearly the entire volcano collapsed. Debris traveled almost 50 kilometers (30 miles) from the volcano in an enormous landslide almost 20 times larger than the Mount St. Helens landslide.

Lahars

Lahar is an Indonesian term for a mixture of water and rock fragments flowing down a volcano or river valley. These flows can rush down valleys and stream channels at speeds of 30 to 60 km per hour (40–50 mph) and can travel more than 60 km (50 miles). As lahars pick up debris and water, they can easily grow to more than 10 times their initial size.

Some lahars contain so much rock debris (60 to 90% by weight) that they look like fast-moving rivers of wet concrete. Close to their source, these flows are powerful enough to rip up and carry trees, houses, and huge boulders miles downstream. As slopes level out and the flows lose energy, they entomb everything in their path in mud.

Historically, lahars have been one of the deadliest volcanic hazards. In 1985, Colombia's Nevado del Ruiz volcano produced a relatively modest VEI 3 eruption. The pyroclastic flow melted about 2.5 sq km (1 sq mi) of snow and glacial ice on the mountain. Melt water mixed with volcanic ash rushed down river valleys at speeds up to 60 km/hr (37 mph), stripping away rocks, soil, and vegetation.

Two hours later, a 5-meter (16-foot) wall of hot mud and debris slammed into the town of Armero, 74 km (46 mi) from the volcano, killing nearly 23,000 people and destroying 5,000 homes. Surprisingly, Armero had previously been devastated by lahars in 1595 and again in 1845, only to be rebuilt on the same spot.

Aerial view of the Armero devastation. The pattern of the city streets and buildings is still visible through the debris. The river valley from which the lahar emerged can be seen in the background.

Volcanic resources

These short-term hazards of volcanoes are balanced by many long-term benefits to humanity. In many places, fertile volcanic soils make it profitable to live near volcanoes. Volcanic materials are used in construction and other industries. Heat energy from young volcanic systems may be harnessed to produce electricity. Most volcanoes are located near oceans, and many are found in mild climates. Because these volcanoes can lie dormant for long periods of time, people often consider them harmless.

The challenge to volcanologists is to minimize the risks associated with volcanic hazards, so that society may continue to enjoy volcanism's long-term benefits. They must continually improve their ability to predict eruptions and provide sound information to decision makers and the public.

Volcanic soils are rich in minerals and have high water-holding capacities. In the state of Washington, volcanic soils around the Cascade mountains, such as Mount Rainier, provide rich cropland.

Climate change

Major eruptions like the June 15, 1991 eruption of Mount Pinatubo inject huge amounts of sulfur dioxide gas (SO_2) into the stratosphere. There, SO_2 combines with water to form tiny droplets of sulfuric acid (H_2SO_4) as shown below. These droplets, called *aerosols*, scatter sunlight, lowering Earth's average surface temperature for long periods of time. The satellite images below show the spread of aerosols from the Pinatubo eruption.

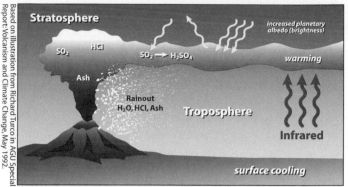

Large eruptions pump fine ash and gases into the stratosphere. There, chemical reactions produce sulfuric acid (H_2SO_4) and other chemicals that increase Earth's albedo, or reflectivity. Less sunlight reaches the ground, causing Earth's surface to get cooler.

For two years after the Pinatubo eruption, global temperatures were about 0.6°C (1.1°F) lower than normal. Sulfuric acid also contributes to the destruction of the ozone layer.

These aerosols may persist for months or years, until they eventually settle out of the atmosphere or are "scrubbed" out by precipitation or other processes. Historically, periods of rapid global cooling have occurred after the largest eruptions.

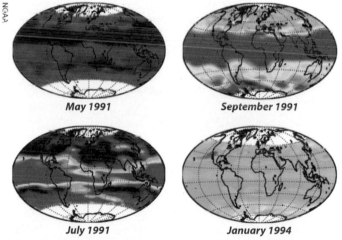

These images show how the transparency of the atmosphere changed following the eruption of Mt. Pinatubo in June 1991. The dark blue colors of the May 1991 image indicate a clear atmosphere. After the eruption, the red band around the equator indicates a "murky" atmosphere that gradually thins as it spreads to higher latitudes.

Questions

1. Why do volcanoes at hot spots erupt less violently than volcanoes near subduction zones?

2. What is tephra, and how does it cause damage?

3. Why is carbon dioxide gas (CO_2) a dangerous eruptive product?

4. In 1985, a relatively mild VEI 3 eruption of Colombia's Nevado del Ruiz volcano destroyed the town of Armero, nearly 50 miles away. What eruptive product or process destroyed the town, and what warnings did they have that this type of event was possible?

5. Describe three ways in which volcanoes benefit humans.

6. How does sulfur dioxide gas from major volcanic eruptions cause global cooling of Earth's climate?

7. How long does the cooling effect from sulfur dioxide gas (SO_2) typically last?

Activity 4.4

Volcanoes and climate

In this part of your investigation of volcanoes, you will look at some of the major eruptions occurring since 1450 AD and explore the relationship between these eruptions and changes in the Earth's climate.

▶ Launch the ArcView GIS application and locate and open the **hazards.apr** file.

▶ Open the **Volcano hazards** view and turn off the **Volcanic Eruptions**, **Volcanoes**, and **Plate Boundaries** themes.

Major Eruptions and the Volcano Explosivity Index (VEI)

How often do major volcanic eruptions occur? You will look at the historical record over the last several hundred years to answer this question.

▶ Turn on and activate the **Major Eruptions** theme. This theme shows the largest volcanic eruptions occurring since 1450 AD.

▶ Click the Query Builder button 🔍 and enter a query to determine the number of major eruptions of VEI category 5. ([VEI] = 5)

To turn a theme on or off, click its checkbox in the Table of Contents.

To activate a theme, click on its name in the Table of Contents. Active themes are indicated by a raised border.

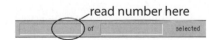

▶ Click the **New Set** button and close the Query Builder window.

▶ Click the Open Theme Table button 🖾 and read the number of major eruptions in the tool bar.

1. Record the number of eruptions of VEI = 5 in the table below.

VEI Value	5	6	7
Number of eruptions of this VEI			
Number of years covered in data set			
Recurrence Interval (average years between eruptions)			

▶ Repeat the query for VEI = 6 and VEI = 7 and record the number of eruptions of each magnitude in the table.

Next you'll determine how many years the data set covers.

▶ Make sure the **Major Eruptions** theme is active. Click on the Open Theme Table button 📖.

▶ Select the Year column and click the Sort Ascending button 📄.

How many years?

To find the number of years covered, subtract the year of the most ancient eruption from the year of the most recent eruption in the data set.

2. How many years are covered by the **Major Eruption** data set? Record this number in the table above. (The number is the same for all three columns.)

The recurrence interval for eruptions of a particular VEI is the average number of years between eruptions of that magnitude.

3. Calculate the VEI recurrence interval by **dividing the number of years by the number of eruptions** in that VEI category. Record the recurrence intervals in the table above.

4. What problem might there be with the recurrence interval you calculated for the largest VEI categories? Does it represent a full recurrence interval?

5. As the VEI increases, what happens to the number of eruptions and the recurrence interval?

Major eruptions and northern hemisphere climate

In this section, you will look at a graph recently published in the science journal *Nature* that shows changes in atmospheric temperature in the northern hemisphere over the last 500 years.

Tree rings record climate

Trees grow by adding layers of new cells to the outer surface. Each spring, as temperatures warm up, trees begin to grow more rapidly. The cells are larger and this new wood appears lighter in color. As temperatures drop in the fall and growth slows, the cells become smaller and the wood appears darker. This pattern repeats itself yearly. By counting the number of these cycles, or rings, you can determine the age of the tree.

The width of each ring is determined by rainfall and other factors. The pattern of ring widths provides a record of the climate during the life of the tree. Matching these patterns between living and dead trees allows scientists to reconstruct long histories of climate for a region.

Scientists can't measure past temperatures directly, but they can infer temperatures indirectly by measuring the spacing of growth rings in trees. Closely spaced rings indicate poor growing conditions such as cold weather or low rainfall, and widely spaced rings indicate warmer temperatures and more rainfall.

The scientists were looking for large *anomalies* in the data, years when the temperature was much lower than normal. You will examine the graph to identify these major atmospheric cooling events, then use Arc-View to determine which volcanic eruption triggered each event.

▶ Click the Media Viewer button 📽 and choose **Volcanoes and Climate** from the list.

The top graph shows how the average temperature varied from normal over a 600 year period. The bottom graph shows the VEI intensity of volcanic eruptions over the same time period. Six temperature anomalies—major atmospheric cooling events—are marked with yellow dots.

6. For each event, read the temperature anomaly on right hand axis of the graph and record it in the table below.

Year of anomaly	Temperature anomaly (° C)	Name of volcano	Date of eruption	VEI of eruption
1453				
1601				
1816-1818				
1884				
1912				
1992				

▶ Close the media viewer window.

▶ Click the Open Theme Table button 🖾 to open the **Major Eruptions** attribute table.

▶ Scroll down the table and look for volcanoes with high VEI values that erupted in the same year as (or the year before) each temperature anomaly you recorded.

7. Use the **Major Eruptions** attribute table to complete the table. Fill in the name, date of eruption, and VEI for each eruption.

8. How well do the years of the temperature anomalies match the dates of the corresponding eruptions? What might account for differences between the two dates?

9. How is the VEI of an eruption related to the degree of atmospheric cooling following the event, as indicated by the temperature anomaly? Give examples to support your answer.

▶ Close the **Major Eruptions** attribute table.

The big ones—prehistoric VEI 7 & 8 eruptions

You've seen that the 1815 Tambora eruption was a big one. The only VEI 7 eruption of the last 550 years, Tambora cooled the Northern Hemisphere by an average of 0.5°C for several years. In fact, 1816 has been called "the year without a summer." The Northeastern US was hit particularly hard, with snow in June and killing frosts from June through September. Beginning in 1817, large numbers of people started leaving northern New England for warmer climates.

The most recent powerful eruption was the VEI 6 eruption of Mt. Pinatubo in the Philippines in 1991. The eruption produced a global cooling of around 0.3 °C. In terms of total material ejected, Tambora was nearly 20 times larger than Pinatubo and about 500 times larger than the 1980 Mt. St. Helens eruption.

▶ Turn off the **Major Eruptions** theme and turn on the **Ashfall Events** and the **Ashfall Sources** themes.

In the not so distant past, geologically speaking, there have been tremendous VEI 8 eruptions that dwarf those of Tambora and Pinatubo. What exactly is the difference between an eruption with a VEI of 7 and one with a VEI of 8? In this next section you will compare some of the larger eruptions in human history with some of the biggest eruptions in *geologic* history by comparing the areas and volumes of their ash fall deposits.

Eastern Hemisphere ashfall deposits

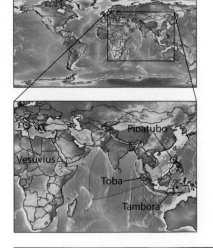

The **Ashfall Events** theme shows the distribution of ash from several historic and prehistoric volcanic eruptions. Notice how the ash plumes overlap, particularly in what is now the United States.

▶ Use the zoom tools to take a closer look at each of the four ashfall plumes in the Eastern Hemisphere (see locator map at left). The volcanoes that produced the plumes are labeled by name. To select the ashfall plumes for Vesuvius and Pinatubo, you will probably have to zoom in even closer.

▶ Open the **Ashfall Events** theme table.

10. Find the VEI and area of each plume and record them in the table below.

▶ Close the **Ashfall Events** theme table.

▶ Use the Measure tool 📐 to find the distance from the source (the red triangle) to the farthest extent of each plume (use your best judgment).

11. Record the distance in the table below. (Round it to the nearest 100 km.)

Volcano name	Eruption VEI	Area of ash fall plume (sq km)	Maximum distance from volcano (km)
data source	*[VEI]*	*[Area (sq km)]*	*measure*
Vesuvius			
Pinatubo			
Tambora			
Toba			

Now, consider the Indonesian volcano Toba, which erupted about 74,000 years ago. Toba deposited ash over an estimated 1% of Earth's surface, left telltale chemical traces in the ice caps of Greenland and Antarctica, and produced a prolonged period of cold and darkness that severely stressed humankind's ancestors. Genetic evidence suggests that the human population may have declined significantly due to Toba's effects, and that our species survived only by finding refuge in isolated pockets of tropical warmth. Scientists estimate that the Toba eruption caused global cooling of 3-5°C (5-9°F) for a period of up to seven years, with summertime temperatures in some areas dropping by as much as 15°C (27°F).

12. Consider the effects that the 0.5°C cooling from the Tambora eruption had on society. How might an eruption similar in magnitude to Toba affect society today?

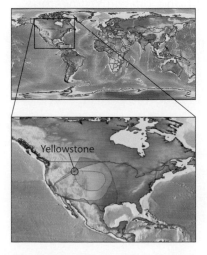

Yellowstone Caldera ashfall deposits

On a human time scale, VEI 8 eruptions are rare, occurring once or twice every 100,000 years. Geologically speaking, however, VEI 8 eruptions are relatively common. Next, you will look at three of the biggest eruptions to occur in the United States over the past 2 million years, the Yellowstone Caldera eruptions.

The Yellowstone Caldera eruptions

Occupying the northwest corner of Wyoming is a spectacular area of geysers, hot springs, and abundant wildlife. This region was so unusual and notable for its geological and biological wonders that it was protected back in 1869 as the United States' first national park. Called Yellowstone, the park today is considered one of the three "crown jewels" of the National Park System, and receives about 3 million visitors per year.

Many visitors are unaware that Yellowstone is a giant collapsed volcano, or caldera, that has produced three gargantuan eruptions within the past 2 million years. Next, you will take a closer look at this slumbering volcanic giant.

▶ Choose **View** ▶ **Full Extent** to view the entire map, then use the Zoom In tool 🔍 to zoom in on the continental United States.

You should be able to see several overlapping ashfall plumes across the middle and southwestern parts of the US. Three of these plumes were produced by volcanic activity at Yellowstone; the fourth plume was created by another massive caldera eruption from the Long Valley Caldera in California.

▶ You will use a theme properties query using a wildcard to display only the Yellowstone ashfall plumes.

- Make sure the **Ashfall Event** theme is active.
- Choose **Theme** ▶ **Properties**.
- Click the Query Builder button 🔧. ← *Don't forget the asterisk (*)!!*
- Enter the query (**[name] = "Yellow*"**) exactly as shown.

Only the Yellowstone ashfall plumes should be visible. Next you will use the ashfall event theme table to obtain information about each eruption.

▶ Click the Open Theme Table button 🗔.

13. Complete the following table with information about the three Yellowstone ashfall deposits. List them in order from oldest to youngest.

Name [Deposit]	Years before present [Age]	Area of deposits (sq km) [Area (sq km)]

14. Estimate the area in square kilometers that the ashfall from the next Yellowstone Caldera eruption might cover. You can do this by calculating the average area of the three historical ashfall deposits.

Finally, calculate the recurrence interval. You can use the recurrence interval to predict when the next major eruption from the Yellowstone Caldera might occur.

- Subtract the youngest age from the oldest age in the table above.
- Divide by the number of times the original eruption was repeated; in this case, divide by 2.

15. What is the recurrence interval for the Yellowstone Caldera?

16. Use the most recent eruption and the recurrence interval to estimate when the next eruption might take place.

Using wildcards in queries

When you're not sure of the exact spelling of a query value or want to find related values, an asterisk (*) can be used to represent any missing characters (or no characters). For example, querying for "Yellow*" would find features containing Yellow, Yellowstone, Yellowjacket, and Yellow No. 2, but not Big Yellow, Mellow Yellow, or The Yellow Brick Road in the queried field.

Unit 5
Tsunami Hazards

In this unit, you will...

- *analyze two major tsunami events in detail,*

- *discover the effects tsunamis have on communities and how communities can prepare for them, and*

- *examine tsunami trigger events and develop criteria for issuing tsunami warnings.*

NOAA/NGDC/Sunset Newspaper

A major earthquake off the coast of Chile on May 22, 1960 produced a tsunami that affected the entire Pacific Basin. In Hilo, Hawaii—10,000 km from the earthquake—the tsunami caused 61 deaths and $24 million in property damage. Frame buildings were either crushed or floated off their foundations and only buildings of reinforced concrete or structural steel remained standing.

Activity 5.1

Scotch Cap Light Station

Location Map

After reading the account of the tsunami that destroyed the Scotch Cap Light Station in 1946 (page 97), list and describe all of the tsunami-related hazards discussed. Feel free to add other hazards from your previous knowledge or experience with tsunamis.

Tsunami hazards

1.

2.

3.

4.

5.

6.

7.

8.

The Scotch Cap Light Station, located on Unimak Island in Alaska, was a five-story reinforced concrete structure built 10 meters (33 feet) above sea level. On April 1, 1946, a strong earthquake jolted the lighthouse. Forty-five minutes later, a tsunami estimated at 35 meters (115 feet) high obliterated the lighthouse.

The eyewitness account reprinted here is the actual report submitted by an officer stationed in the radio shack seen on the cliff above the lighthouse in the bottom picture.

Before (above) and after (below) pictures of the Scotch Cap Lighthouse.

Tsunami hazard examples (optional)

If you have access to a computer, you can see examples of these and other tsunami hazards. Be sure to add any new hazards you find to your list.

▶ Launch **ArcView GIS**.
▶ Choose **File ▶ Open...** and locate and open the **hazards.apr** file.
▶ Open the **Geological Hazards** view.
▶ To see examples of other earthquake hazards:
  • Turn on and activate the **Hazard Links** theme.
  • Using the Hot Link tool 🗲, click on each of the blue tsunami hazard symbols 🌊 (blue waves) on the map.
  • Read the caption for each picture, **then close its window**. There may be more than one picture for each link.

Questions

1. Which of the hazards you listed caused the greatest amount of damage and loss of life at the Scotch Cap light station?

2. Do you think anything could have been done to protect the lighthouse or its staff? Explain.

3. From the account, did the light station staff have warning signs of the coming wave? Explain.

4. Why do you think so much of the world's population lives close to the ocean?

5. Would you rebuild the light station in the same location? Explain.

Eyewitness account
Scotch Cap Light Station

MEMORANDUM KEPT BY CHIEF RADIO ELECTRICIAN
HOBAN B. SANFORD, U. S. COAST GUARD

At 0130 Xray, 1 April, 1946, at which time I was awake and reading, a
severe earthquake was felt. The building (CG Unit 368 - Unimak A/F Station)
creaked and groaned loudly. Objects were shaken from my locker shelves.
Duration of the quake was approximately 30 to 35 seconds. The weather
was clear and calm.

Knowing that the volcanoes to Northward of the building had been active
at one time, I immediately looked in that direction for signs of renewed
activity and upon seeing none made a round of the building to see what,
if any, damage had been caused by the tremor. Inspection failed to reveal
any damage other than objects shaken from locker shelves. The crew were
all awakened by the quake.

Intending to call Scotch Cap Lightstation on the phone to ascertain if
they had felt, or been damaged by, the quake, I went to the phone in
Operations, but Pitts, RM2c, had already done so and he stated they had
felt the tremor and that Pickering, MoMM2c, who was on watch at the
lightstation, had said that he was "plenty scared" and was going to call
Dutch Harbor Navy Radio to see what information that unit might have
regarding the earthquake.

At 0157 Xray a second severe quake was felt. This one was shorter in
duration, lasting approximately 15 to 20 seconds, but harder than at 0130
Xray. I again looked towards the mountains for any signs of volcanic
activity, but still could see none. I made a second round of the building
to see if any damages had resulted but none was apparent.

The crew was gathered in the Recreation Hall discussing the shocks, their
probable cause and location when a crew member stated he had talked with
Scotch Cap Lightstation after the second shock, and they were attempting
to contact Dutch Harbor Radio for any news of the quakes.

At 0218 Xray a terrible roaring sound was heard followed almost immediately by a very heavy blow against the side of the building and about 3
inches of water appeared in the galley, Recreation Hall and passageway.
From the time the noise was heard until the sea struck was a matter of

seconds. I should say between five and ten seconds at most. Ordering the crew to get to the higher ground of the DP D/F building immediately. I went to the control room and, after a couple of calls to Kodiak and Adak Net Control Stations, broadcast a priority message stating we had been struck by a tidal wave and might have to abandon the station, and that I believed Scotch Cap Lightstation was lost.

> (Message: - PPP NMJ NNA NNFV NNBT TIDAL WAVE MAY HAVE TO ABANDON THIS PLACE X BELIEVE NNHX LOST INT R INT R XXX)

Received no answer to calls or receipt for message and did not know until daylight that the receiving antennae had been carried away. Electric power was fluctuating badly and starting for the generator room to ascertain the cause and extent of damage, I found that D'Agostino, ETM1c, and Campanaro, RM2c, had voluntarily remained behind to assist.

Water had struck the switchboards through a burst-in door and the voltage control regulator was burning on the back of the board. ETM1c D'Agostino used a CO_2 extinguisher while I shut down the generator. This placed the station in darkness.

Companaro found and lit a kerosene pressure lantern and we proceeded to make emergency repairs. The switchboards were shunted and the generator connected directly to the line. This restored lighting and some power circuits. Companaro was sent to call back some members of the crew to get more clothing and canned goods to be taken to the DP building in case of a second wave. While crew members were thus engaged, D'Agostino and myself made a rapid survey of damage. At 0345 I went to edge of hill above Scotch Cap Lightstation to observe conditions there. The way was littered with debris, and the lightstation had been completely destroyed. I returned to the DF Station and with D'Agostino, continued cleaning up water and muck about generators. At 0550 had one generator running full power, at this time transmitted a dispatch to the DCGO, 17ND, via Kodiak re conditions. At 0700 went down to the site of the lightstation, the sea by this time having receded to its usual limits, and in company with several crew members searched among the debris for any signs of bodies of personnel. On top of hill behind the lightstation we found a human foot, amputated at the ankle, some small bits of intestine which were apparently from a human being, and what seemed to be a human knee cap. Nothing else was found. At 0725 was informed Sarichef Beacon heard. At 0800 sent out searching parties to attempt locate any trace of Scotch Cap personnel. Searching parties later returned and reported no trace of the lightstation crew. The crew of the lightstation was comprised of Petit, CBM, OinC; Pickering, MoMM2c; Dykstra, S1c; Ness, S1c, and Calvin F1c.

Searching parties were out daily when ever weather permitted until 20 April when CBM Sievers of CGC CLOVER, which was establishing a temporary light on the site of the destroyed light, located a body which was identified as Paul J. Ness, S1c, a member of the lightstation crew. The body was viewed by several crew members and myself and all agreed that it was Ness, who had high cheek bones, slightly prominent upper incisor teeth and a small goatee. The pharmacist mate from Unit 368 had been observing the large toes on both feet of Ness and the nails were pared away from the sides. This condition, also existed on the feet of the body. The remains were wrapped in an old blanket and canvas and removed to above the high water mark, pending burial instructions from DCGO, 17ND. On 22 April at 1030 CBN Sievers; who was conducting a search to eastward, returned to Unit 368 and stated he had found another body. With several crew members I proceeded on to the location but was unable to identify the body. The body was decapitated, disemboweled, and in a poor state of preservation. A home-made monel ring on the right hand could not be identified by any member of the crew of Unit 368.

At 1100 crew members who had been searching to westward reported they had found the right thigh and foot of a man. The foot could not be identified. These remains were gathered in old mail sacks and placed in a rough coffin. The body of Ness was placed in an individual coffin.

At 1545 23 April, the body of Ness was buried in an individual grave, the unidentified portions of bodies were buried in a common grave adjacent thereto. The graves are at the seaward edge of the western bank of the first ravine to the eastward of Scotch Cap Lightstation and are approximately 300 yards from the site of the light, near the graves of two Russian seamen. The graves are plainly marked with white wooden crosses with brass plates securely attached, and are well covered with rocks to discourage depredation by animals.

The area covered by searches was approximately 5 miles eastward, 4 miles westward from Scotch Cap Lightstation, and inland to the high water mark of the tidal wave.

Notes: Most of the abbreviations used in this account are US Coast Guard occupations and ranks: RM2c - Radioman Second Class • MoMM2c - Motor Machinist's Mate Second Class • ETM1c - Electronic Technician's Mate First Class • S1c - Seaman First Class • F1c - Fireman First Class • CBM - Chief Boatswain's Mate • OinC - Officer In Charge. Other abbreviations include the following: DCGO - District Coast Guard Officer • CGC - Coast Guard Cutter. "Xray" is a designation for the time zone covering Unimak Island. It is UTC –11, which means it is 11 hours offset from Greenwich Mean Time. For comparison, The US Eastern time zone is UTC –5, Central is UTC –6, Mountain is UTC –7, and Pacific is UTC –8. "0130 Xray" is thus 7:30 in the morning on the East Coast and 4:30 in the morning on the West Coast.

Source: US Coast Guard

Activity 5.2

Deadly tsunamis

Tsunamis (soo-NAH-meez) are one of the most devastating geologic hazards facing coastal communities. Sometimes mistakenly called "tidal waves," tsunamis are not related to the ocean's tides. Rather, they are caused by the sudden movement of the sea floor or coastline due to geologic events such as earthquakes, volcanic eruptions, or landslides.

When the sea floor moves suddenly, it displaces the overlying water, forming a "hill" or "valley" on the surface. As gravity makes the ocean surface return to its normal level, the energy spreads out rapidly as a series of concentric waves, like ripples on a pond.

Tsunami waves are not like wind-blown ocean waves. Wind-blown waves have amplitudes (heights) up to a few meters and wavelengths (distance from crest to crest) of a few hundred meters. In the open ocean, tsunami wave amplitudes range from a few millimeters to about half a meter depending on the depth of the water, and the tsunamis have wavelengths of hundreds of kilometers.

What does it mean?

The Japanese characters for tsunami translate literally as "harbor wave." Understanding tsunami behavior is important in Japan, where both lives and property have been destroyed by tsunamis throughout history.

津 tsu = *harbor* or *port*
波 nami = *wave*

What's the period of a wave?

The **period** of a wave is the time required for successive wave crests to pass a fixed point.

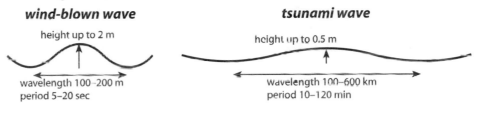

wind-blown wave
height up to 2 m
wavelength 100–200 m
period 5–20 sec

tsunami wave
height up to 0.5 m
wavelength 100–600 km
period 10–120 min

The great tsunami of 1960

▶ Launch **ArcView GIS**.
▶ Choose **File ▶ Open…** and locate and open the **hazards.apr** file.
▶ Open the **1960 Chile Tsunami** view.

On May 22, 1960 a magnitude 9.6 earthquake, the largest ever recorded, struck off the coast of Chile. The location of the quake is marked in this view with a red star symbol.

▶ Turn on the **Plate Boundaries** theme.

1. At which type of plate boundary did the earthquake occur?

▶ Turn off the **Plate Boundaries** theme.

To turn a theme on or off, click its checkbox in the Table of Contents.

The quake generated a huge tsunami that temporarily raised sea level by as much as 30 meters and devastated communities along the Chilean coast. In all, more than 1200 people were killed. Unfortunately, the danger was not limited to the nearby coastline.

In the open ocean, tsunamis are difficult to detect and impossible to see. To study tsunami behavior, researchers build numerical models on computers. One of these models recreates the 1960 Chilean tsunami. The heights of the waves are greatly exaggerated to make them visible.

Where is this?

The animation does not show the entire globe. The star marks the location of the earthquake that produced the tsunami.

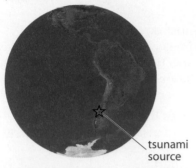

tsunami
source

Animation courtesy of Nobuo Shuto, Disaster Control Research Center, Tohoku University, Japan

The model shows only part of the globe. Use the diagram at left to orient yourself before viewing the animation.

▶ To view the 1960 Chilean tsunami model:
  • Using the Hot Link tool ⚡, click on the tsunami source off the coast of Chile (the red star). Be patient while the movie loads.
  • Watch the movie several times. The waves travel a long distance, so the view rotates twice to show them passing Hawaii and reaching the far side of the Pacific.

After viewing the movie, answer the following questions.

2. How do tsunami waves respond when they strike islands or coastlines?

3. By the time the waves reach Japan, how much energy do they appear to have lost? (All? Most? A little? None?)

Tsunamis exhibit all of the properties of waves. In the animation, notice how the tsunami waves "slosh" back and forth in the ocean as they reflect off coastlines, islands, and shallow underwater landforms. They reflect (bounce) off of obstacles, refract (change direction) as they change speed, and diffract (change direction) as they pass obstacles. As they slosh around the ocean basin, colliding waves may add together or cancel each other out, causing their amplitude to be much higher or lower than expected.

▶ Quit the QuickTime Player application.

Measuring the effects of tsunamis

Often, the first indication of a tsunami is a rapidly falling water level, followed by a rapidly rising water level that inundates low-lying coastal areas. Major tsunamis can also have towering breaker waves. The tsunami's impact on these areas is usually described by two measurements:

  • **run-up** – the maximum height of tsunami waves above normal sea level.
  • **run-in** – the distance the rising water reaches inland from the normal coastline.

Factors that affect run-up

Tsunami run-up and run-in depend on many factors, including the size of the trigger event, the distance traveled, seafloor topography, tides, and the shape of the coastline.

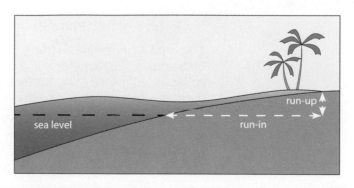

To activate a theme, click on its name in the Table of Contents.

▶ Turn on and activate the **Maximum Run-up** theme.

This theme shows locations that recorded measurable run-up from the 1960 Chilean tsunami. As you saw in the animation, this tsunami affected the entire Pacific basin.

The color of the run-up symbols varies with the height of the run-up. As expected, sites near the source recorded the highest run-up, with the run-up generally decreasing with distance. Unusual circumstances, however, may produce a high run-up far from the source. Next, you will examine the effects of the tsunami on the islands of Hawaii and Japan.

Calculating the tsunami's speed

To calculate the Chilean tsunami's speed, you need to know the distance it traveled and the time it took to go that distance.

▶ Using the Measure tool 📐, measure the distance from the tsunami source to Hawaii. The distance (length) is displayed in the Status Bar, found either above or below the View window.

> Segment Length: 4,622.28 km   Length: 4,622.28 km

4. Record the distance to Hawaii in column 2 of the table below.

location	distance (km)	shortest travel time (hours)	speed (km/hr)	highest run-up (m)
data source ➡	measured	from table	calculated	from table
Hawaii				
Japan				

Measuring tip

Even though the tsunami didn't follow this route, ArcView will correctly measure the shortest distance between any two points.

Measuring the distance from Chile to Japan.

Travel time data format

The travel time data is given in decimal hours, not hours and minutes.

▶ Measure the distance from the tsunami source to the eastern coast of Japan and record it in the table. (See tip at left.)

▶ Find the shortest time required for the leading tsunami wave to reach Hawaii.

• Activate the **Maximum Run-up (m)** theme.

• Zoom in on Hawaii and use the Select Feature tool 🔲 to select the run-up sites in the Hawaiian Islands.

• Choose **Theme ▶ Table...** to open the run-up theme table.

• Scroll across to the **Travel Time** field and click on the field title to select the field. (The title should highlight gray when it is selected.)

• Click the Sort Descending button 📃 to sort the travel times from longest to shortest, then click the Promote button 🖳 to group the sorted Hawaiian run-up sites at the top of the table.

• Scroll through the highlighted items and find the shortest non-zero travel time and record it in the table.

▶ Scroll across to the **Maximum Run-up** field and repeat the sorting and promoting process to find the highest run-up recorded in Hawaii. Record it in the table.

▶ Choose **View** ▶ **Full Extent** and repeat this entire process to find and record the shortest travel time and maximum run-up for sites in Japan.

5. Calculate the average speed of the tsunami between its source and each location using the formula speed = distance / time, and record it in the table.

The average depth of the Pacific Ocean between Hawaii and Japan is greater than it is between Chile and Hawaii.

6. Based on the speeds you calculated, what happens to the speed of a tsunami when the water depth increases?

7. Based on the highest runup recorded at each location, what happens to the energy of the tsunami as it travels away from its source?

*Location of the
1993 Japan Tsunami*

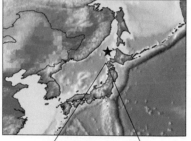

Note: your map does not show the island in this much detail.

Local effects

▶ Close the **1960 Chile Tsunami** view and open the **1993 Japan Tsunami** view.

At 10:17 pm local time on July 12, 1993, a magnitude 7.8 earthquake struck in the Sea of Japan off the coast of Hokkaido, generating one of the worst tsunamis in Japanese history. Run-up reached as high as 30 meters on the nearby island of Okushiri (see locator map at left). In this view, the red star shows the location of the quake's epicenter.

Particularly hard-hit was the resort town of Aonae, on the southern tip of the island, where 185 people lost their lives.

Researchers have modeled the impact of the tsunami on Aonae to help understand what happened and to learn how to better protect coastal communities from damage.

▶ To view an animation of the effects of the tsunami on Aonae:
  • Using the Hot Link tool 🛨, click on the red star symbol. Be patient while the movie loads.
  • The animation shows, from several different angles, the tsunami waves striking the coastline. Watch the movie several times, then answer these questions.

8. From the animation, where do the waves appear to wash the highest and farthest inland? Where do you think are the most dangerous places to live in this area?

▶ Quit the QuickTime Player when you are finished viewing the animation.

▶ To see before and after pictures of damage to the Aonae peninsula, click the Media Viewer button 🖼 and choose **Aonae Before & After** from the media list. The tsunami struck at night, and the after picture shows the area on the following morning.

How much warning?

Aonae locator map

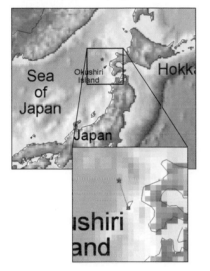

In the 1960 tsunami, it took many hours for the wave to reach Hawaii and Japan. Had an effective warning system been in place, many lives could have been saved. Next, you will find out how much warning the community of Aonae had. Could a warning system have saved the people who died there?

▶ Turn on the **Maximum Run-up** theme, then zoom in on the Sea of Japan until you can see the run-up symbol on the southern tip of the island of Okushiri. (See the locator map at left.)

▶ Using the Measure tool 📏, measure the distance from the tsunami trigger event to the Aonae run-up symbol.

▶ Read the distance from the status bar.

read distance here (your distance will vary)

Segment Length: 1,270.99 km Length: (1,270.99 km)

9. Assuming that the tsunami wave traveled at about 15 km/min (900 km/hr), calculate the travel time to Aonae in minutes using the formula **time = distance / speed**.

10. How much warning did the people of Aonae have? Do you think a tsunami warning system could have saved many lives? Explain.

Activity 5.3

Anatomy of a tsunami

Introduction

Life on Earth evolved in the protective environment of its oceans. To this day, humankind continues to take advantage of the benefits of living along the shore. Moderate temperatures, abundant food, and easy transportation are all provided courtesy of our planet's seas and oceans. Living near the ocean is not without its perils though. Perhaps the most unpredictable and terrifying of these hazards is the tsunami.

A tsunami is a series of large waves created when a disturbance displaces, or moves, an enormous volume of ocean water. Like the ripples that develop when a rock is thrown into a pond, the waves of a tsunami spread outward from their source at very high speed. This means that tsunamis can cross an entire ocean in a matter of hours, making them a truly global hazard for coastal communities.

Tsunamis are a special class of waves called *shallow water waves*. The speed at which they travel is proportional to the depth of the water. In the open ocean, tsunami waves move very fast but have wave heights of 0.5 meter or less. As the water shallows near coastlines, the speed of the wave decreases while its height increases. When the waves reach the shore, they may be tens or, in extreme cases, hundreds of meters high.

Two measurements are used to describe the effect of tsunami waves on the coastline.

- **Run-up** is the maximum height of the tsunami wave above normal tide level.
- **Run-in** is a measure of how far inland the wave reaches beyond the normal shoreline.

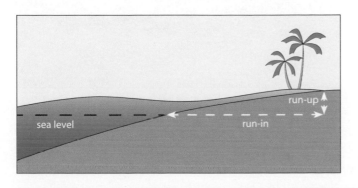

Tsunami magnitude scale

Tsunami researchers have developed a scale, based on the run-up height, for describing the intensity of a tsunami.

Intensity	Run-up Height	Description	Frequency in Pacific Ocean
4	16 m	Disastrous. Near complete destruction of manmade structures.	1 in 10 years
3	8 m	Very large. General flooding, heavy damage to shoreline structures.	1 in 3 years
2	4 m	Large. Flooding of shore, light damage to structures.	1 per year
1	2 m	Moderate. Flooding of gently sloping coasts, slight damage.	1 per 8 months

Tsunami magnitude (similar to earthquake intensity) is a measure of the local size of a tsunami, and is measured with the formula:

$$\text{Magnitude} = 3.32 \times \log(\text{run-up height}).$$

Where tsunamis occur

In the Pacific Ocean, where the majority of tsunamis occur, the historical record shows extensive loss of life and property. Japan, in particular, has repeatedly seen entire towns and cities wiped out by tsunamis, most recently in 1993.

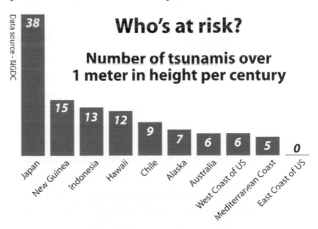

Most coastal communities have some degree of tsunami risk, but the chart shows that tsunamis are far more common in Pacific coastal areas than on the US East Coast. This is because earthquakes and volcanoes, the features that most commonly cause tsunamis, occur more often in the Pacific Basin than in the Atlantic Basin.

Tsunamis of all sizes

Tsunamis occur at various scales, depending on the magnitude of the event that triggers them, the location, and the surrounding topography.

- **Local tsunamis** affect an area within 200 km of their source.
- **Regional tsunamis** affect an area within about 1000 km of their source.
- **Teletsunamis** travel great distances (over 1000 km), often across entire oceans.

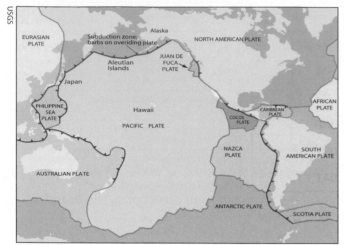

This map explains the prevalence of tsunamis in the Pacific Ocean. The red lines are subduction zones, where one tectonic plate is plunging beneath another. Earthquakes and volcanoes, two common triggers for tsunami events, are typical features of subduction zones.

What causes tsunamis?

There are four types of events capable of producing tsunamis: earthquakes, volcanoes, landslides, and asteroid impacts.

Earthquakes—Normally, only large earthquakes in or near ocean basins produce tsunamis. Of these, only the largest, magnitude 8 and higher, create teletsunamis. However, smaller earthquakes sometimes indirectly cause tsunamis by triggering landslides.

Volcanoes—Some of the most devastating tsunamis in recorded history occurred during the 1883 eruption of the volcano Krakatoa. Tsunami waves with 40-meter run-ups destroyed 165 coastal villages on the Indonesian islands of Java and Sumatra, killing over 36,000 people. The tsunami formed when the volcano either collapsed into its magma chamber or exploded, creating a five-by-nine kilometer "hole." Seawater quickly filled the void, then sloshed outward as an enormous and deadly tsunami.

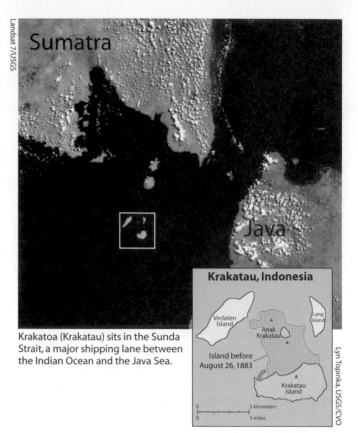

Krakatoa (Krakatau) sits in the Sunda Strait, a major shipping lane between the Indian Ocean and the Java Sea.

Landslides—The movement of rocks and soil can displace large volumes of water, creating a tsunami. These landslides can develop on land and fall into the water or take place completely underwater. One of the largest tsunamis in modern history occurred in Lituya Bay, Alaska, when a magnitude 8.3 earthquake triggered a landslide that fell into an enclosed bay. The slide created a tsunami splash wave that washed 500 meters (1700 feet) over a ridge on the opposite side of the bay.

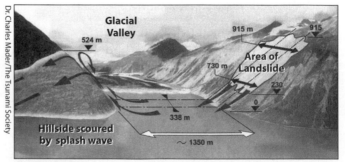

Diagram of the August 9, 1958 Lituya Bay landslide that produced a 524-meter local tsunami wave, the largest in historical times.

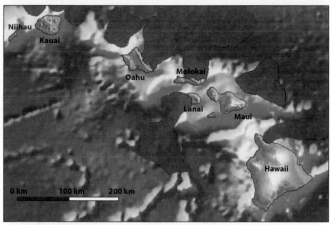

Locations of major Hawaiian submarine landslides. Similar landslides may threaten the northwest and eastern US coasts with tsunamis.

Scientists are also investigating submarine landslides for their tsunami-causing potential. Far more massive than terrestrial landslides, underwater landslides off the Hawaiian Islands have sent thousands of cubic kilometers of material sliding almost 200 kilometers from their source.

The resulting tsunamis would have been enormous. There is speculation today about the origin of coral boulders 200 meters above sea level on the Hawaiian Island of Molokai. Were they tossed there by a massive tsunami, or deposited by "normal" processes when the sea level was significantly higher?

Asteroid and comet impacts—The rarest yet most catastrophic cause of tsunamis is an asteroid or comet impact in one of Earth's oceans. Scientists estimate that a 1-kilometer diameter asteroid striking the middle of an ocean would produce a tsunami with run-ups ranging from 6 to 50 meters.

The asteroid that hit near Mexico's Yucatan Peninsula 65 million years ago, contributing to the extinction of the dinosaurs, was about 10 kilometers in diameter. Evidence suggests that the impact generated a tsunami 100 to 250 meters high that washed hundreds of kilometers inland!

Record of a tsunami event

A tsunami is not a single wave, but a series of wave build-ups and retreats. The waves may be spaced minutes or hours apart. People often believe that the tsunami danger is over after the first wave has passed and are then killed when they return to coastal areas to clean up after the event.

To complicate matters, tsunami waves reflect off coastlines and may pass a given location several times and from different directions. The highest run-up can occur several hours after the arrival of the first wave. Tides are another key factor, as run-up is more severe when it occurs at high tide than at low tide. All of these features of a tsunami event can be seen in the tide gauge record from the 1960 Chilean tsunami shown below.

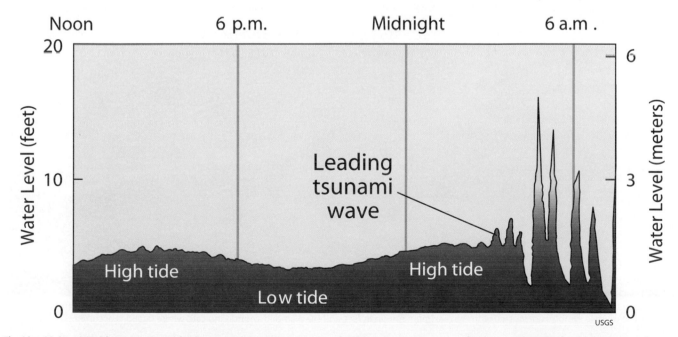

The May 23-24, 1960 tide gauge record for Onagawa, Japan shows a series of eight tsunami waves over a five hour period. The first waves are not the highest. Note the deep trough preceding the first large tsunami wave.

Tsunami effects

Direct Impact—Most of a tsunami's initial damage comes from the direct impact of waves on structures and harbor facilities, and from wave run-up on coastal buildings.

To get a better idea of the damage a large tsunami wave can do, consider that a cubic meter of water has a mass of 1000 kg (2200 lbs). Therefore, a 15-meter high tsunami with a wavelength of 300 meters would hit a 30-meter length of seawall with a 68 million kilogram (150 million pounds) wedge of water—enough to do incredible damage. Debris caught in the backwash of the leading wave makes the secondary waves even more destructive as they come ashore.

Flooding—In addition to the destructive force of the tsunami waves, flooding can kill people, damage property, and spread pollution over a large area.

As shown in the photograph below, run-in from the 1960 Chilean tsunami extended about 3.2 km (2 miles) inland over flat coastal areas and penetrated almost 8 km (5 miles) up the river channels. The run-in is influenced by many factors including the run-up height, orientation relative to the coast, tidal conditions, local topography, and vegetation.

Preparing for tsunamis

Although we can't prevent natural disasters from occurring, we can reduce their damaging effects through effective planning. Scientists and planners are working to improve our ability to detect tsunamis and issue accurate and timely warnings, to respond appropriately when they occur, and to avoid hazardous situations wherever possible.

Planning ahead

Using historical and geological records, planners can predict where and how often tsunami trigger events are likely to occur. Simulating these events using mathematical models allows them to "see" the effects of tsunamis before they occur and to take corrective measures. Many coastal communities are developing tsunami preparedness plans using inundation models like the one shown below. Public education programs are a big part of these plans. For example, some communities now use special signs to alert the public to tsunami hazard zones and evacuation routes.

When high-risk areas are identified before they are developed, appropriate zoning laws ensure that

Markings on this 1944 photograph of the area surrounding the mouth of Río Maullín on the Chilean coast show the extent of run-in from the 1960 tsunami. The triangle symbols are labeled with the maximum run-up, in feet, and the diamond symbols indicate where fatalities occurred.

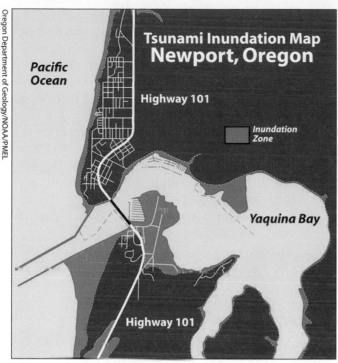

Emergency managers use mathematical modeling to create tsunami inundation maps such as this. Areas prone to flooding are shown in orange, and roads and highways used as evacuation routes are in white.

residential developments and large construction projects such as power stations are restricted to higher ground. In areas that have already been developed, retaining walls can be built to provide a higher level of protection. Following the 1993 tsunami, Japan's Okushiri Island built a 15-meter reinforced concrete wall to protect vulnerable areas.

Monitoring and warning

Today, Pacific Rim countries use a combination of technology and international cooperation to detect tsunamis. The center of operations for this system is the Pacific Tsunami Warning Center located in Ewa Beach, Hawaii. The Center's objectives are to detect and pinpoint major earthquakes in the Pacific region, determine whether they have generated a tsunami, and provide timely warnings to people living in the Pacific region.

Deep ocean tsunami waves are difficult to detect. They average less than half a meter high, have wavelengths of hundreds of kilometers, and move at hundreds of kilometers per hour. In 1995, the Pacific Marine Environment Lab (PMEL) developed a system called DART, short for Deep-ocean Assessment and Reporting of Tsunamis. This system uses sensitive detectors to measure water pressure changes from passing tsunami waves.

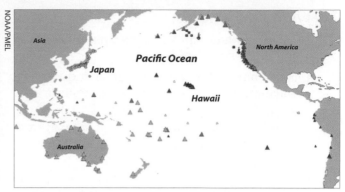

The Pacific tsunami detection system uses twenty-four seismic stations, fifty-three tide stations, fifty-two dissemination points, and six DART systems scattered throughout the Pacific Basin.

They are capable of detecting deep ocean tsunamis with amplitudes as small as 1 centimeter.

Large tsunamis are rare, and developing an accurate warning system is a challenging goal. Historically, nearly 75% of tsunami warnings have been false alarms. For this reason, people are often hesitant to evacuate their homes and businesses, and their response to warnings in general is poor. Emergency managers in tsunami-prone areas must work constantly to increase public awareness and acceptance of the risks of tsunamis and the emergency plans that are in place.

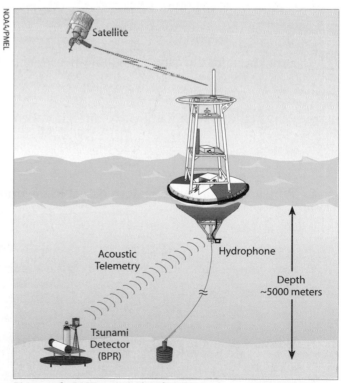

Diagram of a DART station. Placed on the sea floor, a Bottom Pressure Recorder (BPR) transmits pressure information to a surface buoy. The buoy then sends the data to the warning center through a satellite link.

Surviving a tsunami

By interviewing tsunami survivors, planners have put together these survival tips:

- **Heed natural warnings.** An earthquake is a warning that a tsunami may be coming, as is a rapid fall or rise of sea level.
- **Heed official warnings.** Play it safe, even if there have been false alarms in the past or you think the danger has passed.
- **Expect many waves.** The next wave may be bigger, and the tsunami may last for hours.
- **Head for high ground and stay there.** Move uphill or inland, away from the coast.
- **Abandon your belongings.** Save your life, not your possessions.

- **Don't count on the roads.** When fleeing a tsunami, roads may be jammed, blocked, or damaged.
- **Go to an upper floor or roof of a building.** If you are trapped and unable to reach high ground, go to an upper story of a sturdy building or get on its roof.
- **Climb a tree.** As a last resort, if you're trapped on low ground, climb a strong tree.
- **Climb onto something that floats.** If you are swept up by a tsunami, look for something to use as a raft.
- **Expect company.** Be prepared to shelter your neighbors.

Excerpted from U.S. Geological Survey Circular 1187.

Questions

1. What country is most at risk from tsunamis? Why is this?

2. Why doesn't the East Coast of the US experience tsunamis more often? Are tsunamis possible there?

3. Over geological time, what has caused the most destructive tsunamis in the Hawaiian Islands?

4. How often do devastating tsunamis with run-ups of 15 meters or more occur in the Pacific Basin?

5. Why is the first tsunami wave often not the most dangerous?

6. Why are tsunamis difficult to detect in the open ocean?

7. How can a community be prepared for a tsunami?

8. When is it safe to return to coastal areas after a tsunami?

Activity 5.4

Tsunami warning

The 1964 Alaskan tsunami

▶ Launch the ArcView GIS application then locate and open the **hazards.apr** file.

▶ Open the **1964 Alaska Tsunami** view.

The 1964 Alaska earthquake was the second largest earthquake in recorded history and the largest to strike US territory. The earthquake, shown with the red star symbol, caused extensive damage to Anchorage both through shaking and liquefaction of the soil. The earthquake also triggered a major tsunami that seriously impacted many coastal communities, causing fatalities as far away as Eureka, California.

▶ Turn on the **Plate Boundaries** theme.

Although the epicenter of the earthquake was inland, the greatest motion of the seafloor took place some distance offshore, along the Aleutian Trench. An 800-kilometer long slab of the North American plate was thrust suddenly upward by as much as 40 meters as the Pacific plate plunged beneath it.

▶ Turn on the **Tsunami Source** theme. This shows the approximate extent of the displaced sea floor.

▶ Turn on the **Maximum Run-up** theme. This theme shows sites that recorded one or more run-up measurements from the tsunami.

▶ To view an animation of the effects of the 1964 Alaska tsunami:
- Activate the **Trigger Event** theme and click on the trigger event symbol (the red star) using the Hot Link tool ⚡. Be patient while the movie loads.
- Use the forward and backward arrow keys on your keyboard to examine the movie one frame at a time.

To turn a theme on or off, click its checkbox in the Table of Contents.

To use the Hot Link tool, position the *tip* of the lightning bolt cursor over the feature and click.

Like this   Not like this

This movie is different from the others you've seen. Rather than show a simulated wave, this movie uses colors to show changes from average sea level. Light blues, oranges, and reds are higher than normal sea level, and darker blues are lower than normal.

The digital clock at the bottom of the movie shows the number of hours and minutes that have passed since the trigger event. The movie begins twenty-seven minutes before the earthquake and ends almost fifteen hours later.

Use the arrow keys to go to the frame where the clock reads 00:03. Notice the long red-orange "hill" of water that forms off the Alaska coast. As you advance the movie forward in time, the hill spreads out. You can follow the light blue leading edge of the tsunami southward. (Shown as a white dashed line in the illustration at left.)

Tracking the leading wave

At 1:09 after the earthquake event, the leading tsunami wave appears as a light blue arc traveling southward from the source. (The dashed white line does not appear in the movie.)

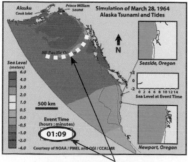

leading wave at 1:09

Tide gauge plots

Tide gauge plots show the local water level at the gauge as it changes over time. A typical plot, in the absence of factors such as tsunamis, shows a gentle pattern of high and low tides that repeats with a period of around 13 hours.

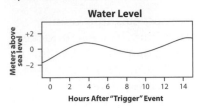

This plot shows what the Seaside, Oregon tide gauge might have registered if the 1964 Alaska tsunami had never happened. The starting point and time scale of the plot are the same as the graph in the movie.

Tides and tsunamis

Tide gauges are devices used to record changes in sea level at coastal locations. Run the movie through several more times, this time focusing your attention on the tide gauge record at Seaside, Oregon. Normal tides would create a regular, gently changing record of the water level like the one shown at left. Watch the leading wave and the tide gauge; you should be able to see the "arrival signature" of the leading wave on the tide gauge readout.

- The arrival signature is found where the gauge readout in the movie first differs from the plot in the sidebar to the left!

1. How long does it take the leading wave to reach Seaside, Oregon?

Continue the movie and watch the tide gauge. Earlier you read that tsunami waves slosh around an ocean basin like water in a bathtub. If the waves have enough energy, they can reflect off of coastlines and return at a later time. When you reach the end of the movie, answer the following questions.

2. According to the tide gauge, how many secondary tsunami waves strike Seaside after the leading wave? (One wave may be very difficult to see in the plot!)

3. Is the leading wave the highest? If not, how many hours after the trigger event does the highest sea level run-up occur? There were actually two peak waves at Seaside; record them both.

4. Based on what you've seen, is it safe to return to low-lying coastal areas immediately after the leading wave? Explain your answer.

Tsunami trigger events

▶ Close the **1964 Alaska Tsunami** view and open the **Tsunami Hazards** view.

Now that you have an idea what tsunamis are and how they travel, you will look at the geologic "trigger" events that cause them. Any event that displaces a large volume of seawater can generate a tsunami. These trigger events include volcanic eruptions, underwater and coastal landslides, earthquakes, and even (rarely, thank goodness) asteroid impacts. Each of the red star symbols in the **Tsunami Sources** theme represents an event that triggered a tsunami in the 20th century. Next, you will examine these data to find out how common each type of trigger event is.

How often do tsunamis occur?

▶ Click the Open Theme Table button  to open the tsunami source theme table.

▶ Read the total number of events recorded since 1900 on the tool bar.

read total here

5. What is the average number of tsunami events recorded each year? (Divide the total number of events by 100 years.)

What causes most tsunamis?

▶ Scroll across the table to the **Trigger Event** field. Scroll down the table to see the different types of trigger events.

6. What is the most common type of tsunami trigger event?

▶ Close the theme table window.

Tsunami warnings

Not all earthquakes produce tsunamis. It's a good idea to warn people of an approaching tsunami, but evacuations are expensive and carry their own risks (panic, looting, etc.). Is there some minimum earthquake magnitude associated with tsunamis? To find out, you will query the Tsunami Sources theme for all events that have an earthquake magnitude listed (i.e., a magnitude > 0).

▶ Build a query to find tsunami sources with listed earthquake magnitudes.

- Activate the **Tsunami Sources** theme.
- Click the Query Builder  button.
- Enter the query (**[EQ Mag (ms)] > 0**) as shown below, click the **New Set** button, and close the query builder window.

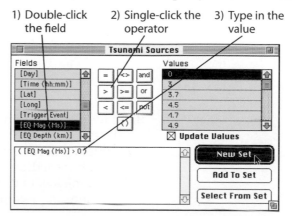

The tsunamis generated by earthquakes with known magnitudes are now highlighted yellow.

▶ Click the Open Theme Table button  to open the tsunami sources theme table.

To activate a theme, click on its name in the Table of Contents.

▸ Find magnitude statistics for these tsunamis:
- Scroll the table to the **EQ Mag (ms)** field heading and click on the heading to select it. (The field name is shaded when it is selected.)
- Choose **Field ▸ Statistics** to display statistics for the field. (**Min** is the lowest magnitude, **Max** is the highest, and **Mean** is the average.)

7. Record the following statistics about the earthquakes that cause tsunamis.
   a. Average magnitude (**Mean**) =
   b. Highest magnitude (**Max**) =
   c. Lowest magnitude (**Min**) =

Use these statistics to help you answer the following questions.

8. Using what you have learned, create a list of criteria that you would use to decide whether and when to issue a tsunami warning. Explain each of your criteria.
   a. What size trigger event requires a warning? How close or how far away would it have to occur?

   b. How would your local geography figure into your decision?

   c. When would you issue the warning?

   d. Who would you notify? How would you notify them?

   e. What would you tell them?

   f. When would you issue an "all clear" signal?

9. Since 1948, more than 75% of tsunami warnings have been "false alarms" because it is difficult to predict the impact of a tsunami. Currently a warning is issued each time there is a magnitude 6.7 or higher earthquake near a coastline or in the open ocean. Do you think it is better to "assume the worst" and send out too many warnings or "assume the best" and sent out too few? Explain.